THE NUT BUTTER COOKBOOK

quadrille

THE NUT BUTTER COOKBOOK

Pippa Murray

Photography by Adrian Lawrence

Publishing Director: Sarah Lavelle
Creative Director: Helen Lewis
Design Concept: B&B Studio www.bandb-studio.co.uk
Senior Designer: Nicola Ellis
Assistant Designer: Gemma Hayden
Editorial Assistant: Harriet Butt
Photographer: Adrian Lawrence
Food Stylist: Emily Kydd
Prop Stylist: Luis Peral-Aranda
Production Director: Vincent Smith
Production Controller: Emily Noto

First published in 2016 by
Quadrille Publishing Limited
Pentagon House
52-54 Southwark Street
London SE1 1UN
www.quadrille.co.uk

Quadrille is an imprint of Hardie Grant
www.hardiegrant.com.au

Text © 2016 Pippa Murray
Photography © 2016 Adrian Lawrence

Cataloguing in Publication Data: a catalogue
record for this book is available from the
British Library.

ISBN: 978 184949 985 9

Printed in China

CONTENTS

1. intro

uction

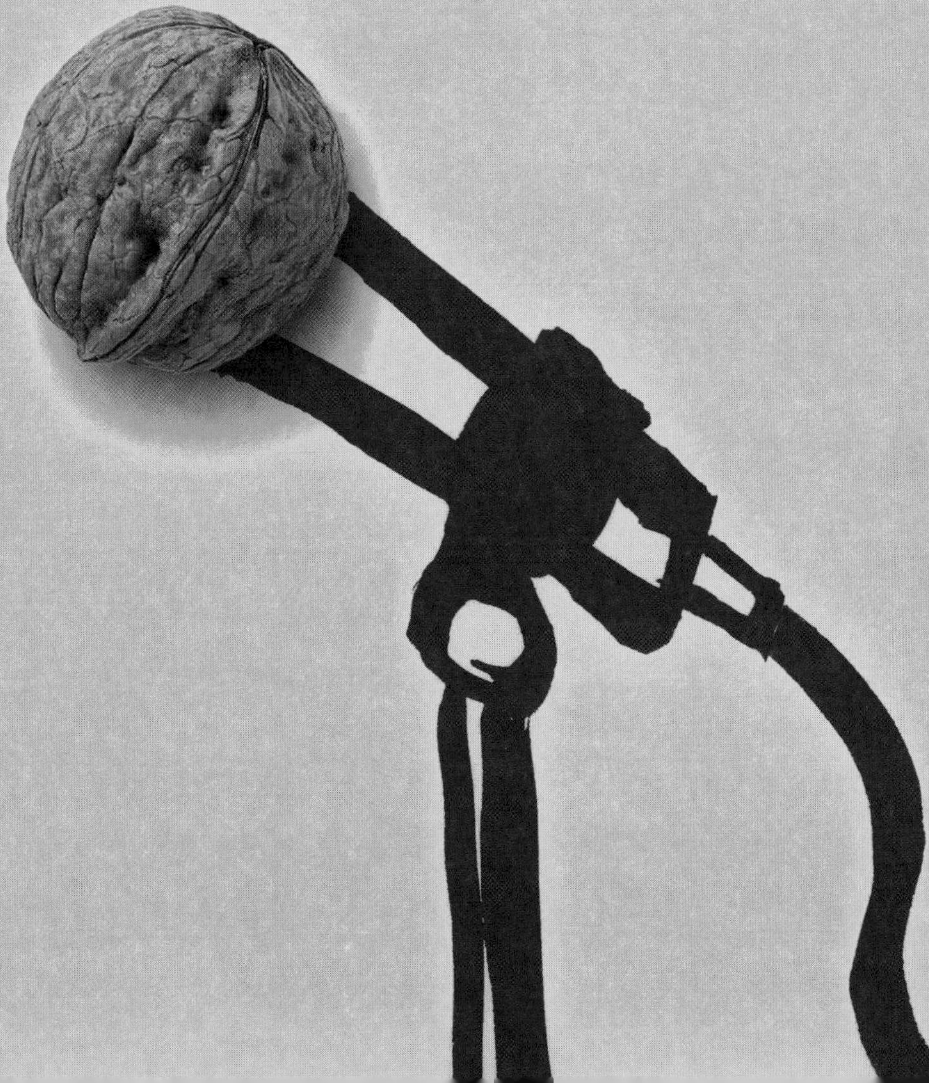

HELLO
FROM PIP

Hi! My name's Pip and I'm the founder of Pip & Nut.

The observant readers amongst you may have guessed that Pip & Nut make nut butter. Not just any nut butter, mind. Our nut butter is 100% natural, free of nasty things like palm oil and sugars, and tastes bloomin' amazing. What started a few years ago as just a twinkle in my eye, a few bags of nuts and a pretty swish blender in my kitchen in North London, has grown into a brand that is now being stocked in lots of shops around the UK and Ireland, and has now given me the opportunity to write this recipe book.

But what spurred me to start Pip & Nut? Well, it came about through my combined love of two things: running and nut butter. Bit random, I know, so let me explain the connection. I spend a large amount of my free time running, sometimes short runs, other times marathons, and I like eating nut butter as my post-run reward. Not only do I love the stuff, but the fact that it is high in protein means that it's pretty good at giving me a bit more energy to help me run that little bit faster. Or at least that's my excuse. However, when looking at the products available to me in supermarkets I noticed many had palm oils and sugars in there. Not being that happy about this, I decided it was time to bring something better to the table. Literally.

I got my blender out, played around with different nuts and flavours and started taking my kitchen-table experiments to markets at weekends. Lots of samples and satisfied customers later, I decided to take the plunge and scale up to get the products to more people. And the rest, as they say, is history.

Whilst Pip & Nut is a company that sets out, first and foremost, to bring you delicious nut butter (you're welcome), the founding principles behind our company go a little bit deeper than that. We're big believers that food, and in particular healthy food, should never, ever be boring. You shouldn't feel like you've made a sacrifice by choosing a healthier option. Food should be a joy to make and then eat.

That's why I'm super-proud of this book. It stands for the fact that from a healthy product – nut butter – you can create incredibly tasty dishes. Some of these recipes you could say are pretty 'good', while others you'd probably class as a bit 'naughty'; but you can safely say that you'll never describe them as bland.

And yes, whilst nut butter as an ingredient is healthy – it's just nuts, after all – there are a few key things I hope you notice when flicking through this book: that nut butter is really versatile and that you can have a lot of fun experimenting with it; that there's more to nut butter than just spreading some on a slice of plain white toast for breakfast; that actually you can pep up your morning smoothie before work, add it to your weekend baking or use it to give your evening meal a boost.

So what are you waiting for? Get cracking (I had to get a nut pun in there somewhere. It'll be the only one I promise!).

IT'S THE NUTS

ALMONDS

MACADAMIA NUTS

WALNUTS

HAZELNUTS

PECANS

PEANUTS

PINE NUTS

CASHEW NUTS

BRAZIL NUTS

PISTACHIOS

IT'S THE NUTS

It's not often that nuts get attention. Perhaps it's because their more colourful friends – fruit – take all the limelight. But seeing as this is a nut butter cookbook, we didn't think we could start without giving you the low-down on nuts. So, bananas, mangoes, apples, raspberries: move aside. It's time to talk nuts.

There are a lot of different types of nuts out there and also lots of nuts that are masquerading as seeds*. Below you'll find a run through of the key nuts that crop up in this book, a bit of information about the nutritional benefits behind them and a fact for you to share with your friends down the pub. They'll be impressed. Well, kind of.

Without making things too complicated we've asterisked the ones that claim they are a nut but are actually a seed or legume!

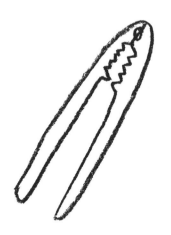

LEBANESE PROVERB: "NOT EVERYTHING ROUND IS A NUT, NOT EVERYTHING LONG IS A BANANA."

ALMONDS

Where are they grown? The fact that over 80% of the world's almonds come from sunny California means they're a pretty big deal out there. So big they've got their very own association, "Almond Board of California". Fancy!

Almonds are harvested once a year in California, and if you ever want to go and see something really beautiful, then head to Cali in February to March, when you'll see Sacramento covered in light pink and white blossoms, the first stage of the cycle of almond crops. Bees then get busy pollinating the trees to kick-start the growth of the crop. Six months later these almonds will be ready to be shaken from their trees and harvested.

What's so special about them? People get excited about almonds because they are packed with lots of minerals, like magnesium, and vitamins, like vitamin B2, which make your hair shiny and nails strong. They also contain naturally high levels of protein. This, coupled with their popularity in almond oil and milk, makes them the UK's second favourite nut. Nice.

Fact! 40% of the world's almonds are bought by chocolate manufacturers – surprised? Blame it on all those Valentine's Day assortments.

BRAZIL NUTS*

Where are they grown? Funnily enough you can find Brazil nuts growing in Brazil. But Brazil isn't the world's biggest producer of Brazil nuts – that would be Bolivia, which produces about 50% of the world's supply. "Bolivia nuts" doesn't have quite the same ring to it though.

What's so special about them? It's no small feat to grow these nuts. They grow near the tops of 150-foot trees in hard casings similar to coconuts. Each case has 20 to 30 nuts snuggled inside, arranged like the segments of an orange. Brazil nut trees are quite particular, too. They require a specific bee for pollination and can take as long as 30 years to mature.

Fact! The cases fall off the trees when ripe, and are easily heavy enough to kill a person.

CASHEW NUTS*

Where are they grown? The cashew tree is native to Brazil's Amazon rainforest. It was spread all over the planet by Portuguese explorers, and today is cultivated on a commercial scale in Brazil, Vietnam, India and many African countries.

What's so special about them? A cashew tree bears numerous, edible, pear-shaped "false fruits", called cashew apples, and on the bottom you'll see the cashew nut. But here's the catch: the nut is encased in a poisonous shell! Which is why, for safe eating, they must go through a rigorous roasting or steaming process to remove the substance.

Nutritionally, cashews have a lower fat content than most other nuts.

Fact! The cashew apple has several culinary uses in its native countries, but because the skin of this fruit is very delicate, it can't be exported. This is why most of us have never seen a cashew apple before!

COBNUTS & HAZELNUTS

Where are they grown? Hazelnuts and cobnuts are one and the same. Hazelnuts are primarily found in Iran and Turkey, while in Britain we grow cobnuts. Whilst hazelnuts are dried, cobnuts are unique in that they are typically sold fresh, which gives the nuts a seasonal market and unique culinary uses.

But beware: if you are hoping to get your hands on these nuts you will have to deal with your main competitor – grey squirrels (pesky little creatures). They will strip a tree in no time.

What's so special about them? With the sole exception of almonds, this is the nut with the highest content of vitamin E. Hazelnuts have become pretty inseparable from chocolate and are most famously used to make praline and … *drum roll* … Nutella!

Fact! Ancient Greeks believed hazelnuts could treat coughing and baldness.

MACADAMIA

Where are they grown? Hawaii might not spring to mind when you think of macadamia nuts, but this small archipelago in the Central Pacific is the largest exporter in the world.

What's so special about them? They contain high amounts of vitamin B1 and magnesium. Besides having a great omega-3 to omega-6 ratio – which helps in fighting inflammation – macadamia nuts contain the largest amount of monounsaturated fatty acids of any nut.

But watch out: Never feed macadamia nuts to your four-legged best friend – they're toxic to canines!

Fact! They have a National Macadamia Nut Day in the USA on 4th September – that's dedication.

PEANUTS*

Hands down the most popular and beloved nut in the world, beating the almond to first place. While "nut" is in their name, peanuts are in fact legumes.

Where are they grown? Two-thirds of globally produced peanuts originate in China and India, but Argentina and the USA are big growers too.

What's so special about them? The beneficial plant fat in peanuts, which is about 80% unsaturated (considered the "good" fat – see below) can help lower cholesterol levels when it replaces saturated animal fat in the diet. Peanuts and peanut butter are also naturally cholesterol-free!

Fact! For those of you who have an irrational fear of getting peanut butter stuck to the roof of your mouth, then clearly you're not alone. This can be diagnosed as Arachibutyrophobia.

PECANS

Where are they grown? Pecan trees are native to North America, and over 80% of the world's pecan crop comes from the USA.

What's so special about them? There are more than a whopping 1,000 varieties of pecan. Many are named after Native American Indian tribes, including Cheyenne, Mohawk, Sioux, Choctaw and Shawnee.

Fact! About 78 pecans are used in the making of an average pecan pie.

PINE NUTS*

Where are they grown? Crunchy yet buttery in texture, pine nuts are pleasantly sweet. They are the small edible seeds of the female cone on the pine trees. While all pine trees will produce a pine nut, only 18 species in the world actually produce large enough nuts to be eaten. These trees are found across Asia, Europe and North America.

What's so special about them? Pine nuts are gluten-free tree nuts, but are probably best known for their use in pesto. It may surprise you to learn that pine nuts can be a potent appetite suppressor. Why? They're a good source of a polyunsaturated fat.

Fact! Apparently, the ancient Greeks and Romans also ate pine nuts. Archaeologists have found the seeds in the ruins of Pompeii.

PISTACHIOS

Where are they grown? 90% of all pistachios are grown in Turkey and Iran.

What's so special about them? If you happen to be feeling stressed, then eat a handful of pistachios. They have a significant amount of potassium that helps in lowering the stress hormone cortisol in our body. Pistachios are also called skinny nuts – one pistachio nut has just 3 calories!

Fact! The country that consumes the most pistachios is China, totalling an impressive 80,000 tons a year – that's the equivalent of 28,000 elephants or eight Eiffel towers!

WALNUTS*

Where are they grown? Walnuts are the oldest known tree food, dating all the way back to 10,000BC. They were brought to California in the 18th century, and this US state now produces 75% of the world's supply of walnuts. No wonder they are so wrinkly!

What's so special about them? Due to their appearance, with the shell shaped like a human skull and the kernel resembling a brain, walnuts have always been regarded as "brain food". Recent studies have shown that they do indeed promote brain function because of their omega-3 fatty acid content.

Fact! Because walnuts resemble the brain, they were believed in medieval times to be able to cure headaches.

IT'S THE ~~NUTS~~ SEEDS

We know this is a nut butter book so talking about seeds shouldn't technically be on the agenda, but they're actually quite closely linked. All seeds are encased in some form of fruit. Fruit is anything that protects a seed. Nuts are a type of fruit, and they also encase seeds. Confused? We don't blame you. Well, the long and short of it is that as you can also make some pretty tasty recipes using seed butter, and as it's a fairly similar process to making nut butter, they are worth a mention. Here's a bit of information on some of the seeds we use throughout the book.

PUMPKIN SEEDS

Also known as pepitas for those who speak Spanish, these are flat, dark green seeds that you can sometimes find encased in a yellow-white husk.

Where are they grown? Today, China produces more pumpkins and pumpkin seeds than any other country, where they are traditionally eaten as a treat at small get-togethers, especially when friends are gathered to chat over a cup of tea.

What's so special about them? While antioxidant nutrients are found in most nuts and seeds, it's the diversity of antioxidants in pumpkin seeds that makes them unique in their antioxidant support. Pumpkin seeds contain vitamin E in its very wide range of varieties: alpha-tocopherol, gamma-tocopherol, delta-tocopherol, alpha-tocomonoenol and gamma-tocomonoenol. The list goes on, but we don't want to send you to sleep!

Fact! A quick Google search reveals that each pumpkin has about 500 seeds – we'll take their word on that one.

SESAME SEEDS

Where are they grown? The sesame plant is a tall annual herb that belongs to the rather fancy Pedaliacea family, and grows extensively in Asia, particularly in Burma, China and India.

What's so special about them? Sesame seeds may be tiny, but boy do they pack a punch when it comes to health benefits. In fact they were worth their weight in gold during the Middle Ages, and for many good reasons. These little guys are notably high in zinc, copper, manganese, calcium and magnesium, giving your skin elasticity and glow, your bones strength, and lowering your blood pressure!

Fact! Sesame helps protect you from the impact of alcohol on your liver, contributing to healthy liver function (although we're not suggesting that eating a handful cancels out that extra glass of wine!).

SUNFLOWER SEEDS

Where are they grown? Sunflower seeds have been knocking around the block for a while. They were one of the first plants ever to be cultivated in the United States and have been used for more than 5,000 years by the Native Americans.

What's so special about them? Sunflower seeds aren't just pretty faces; sunflowers are actually good at absorbing toxins, too. Hence planting sunflowers can help soak up nuclear radiation!

Sunflower seeds are an excellent source of vitamin E and a very good source of copper and vitamin B1.

Fact! Shelled sunflower seeds are one of the most popular snacks in Russia – you can buy them outside metro stations in small newspaper cones.

SESAME SEEDS

PUMPKIN SEEDS

P-P-P PROTEIN

Now, protein has certainly been enjoying its time in the limelight recently. But what's all the fuss about? In short, protein isn't just essential for a healthy diet; it is essential for a healthy life. Protein is found throughout the body: in muscle, bone, skin, hair and virtually every other body part or tissue. There are literally thousands of different proteins that make us who we are. In fact, the only other substance more plentiful than protein in the body is water. But what does protein actually do? We need protein to grow, heal and carry out almost every chemical reaction in the body. Basically, we need it to live.

SO, WHERE CAN I FIND PROTEIN?

Like simple and complex carbohydrates, proteins are absorbed at different rates in the body. Whey protein is known for being quick to digest and this is why many people choose to have it before or after exercising – helping those muscles to recover and grow. On the other hand, casein, the primary protein in milk, releases its amino acids slowly. This means it is particularly beneficial when consumed in the morning, between meals and at bedtime. There are a lot of protein powders/bars/balls (just about anything!) on the market that are designed to be a quick and easy way to get protein into your body, particularly for people who work out a lot. However, we shouldn't forget that there are some fantastic natural sources of protein too, like eggs, fish, meat and, of course, nuts.

Ok, so we know this is a nut butter cookbook and nuts are our "thing", but they really are a fantastic source of natural protein too – promise! Obviously, different nuts have different nutritional values and so some contain more protein than others. Almonds rank up there near the top, with cashews, pistachios and peanuts also packing a healthy punch of protein.

HOW MUCH PROTEIN DO I NEED IN MY DIET?

This is a bit of a tricky question as it really depends on how active you are, as well as a few other factors. Basically, we need a small amount of protein to survive, but we need a lot more to thrive. Based on the Reference Nutrient Intake (RNI), adults in the UK are advised to eat 0.75g protein for each 1kg/2.2lb they weigh. However, many people eat significantly more than this, especially if they exercise frequently. That being said, there are many medical experts who warn against excessive consumption of protein – we told you it was a minefield – so it's important to judge your intake against your own body and the amount of activity you do.

GOOD FATS vs BAD FATS

Let's talk about fat, baby*
Let's talk about you and me
Let's talk about all the good things
And the bad things that may be

adapted from 1990s pop sensation Salt-N-Pepa

Being up-front and honest, nuts contain fat. STOP. Don't throw the book in the bin. Let us explain. There are good and bad things about fats (to paraphrase Salt-N-Pepa).

Yes, some "bad" fats – eaten in excess – can be found guilty of giving you a shock when you stand on the bathroom scales, but "good" fats, like omega-3, can actually help fight fatigue, keep you in a good mood and improve your brain power.

Now, without getting too technical (nobody wants a science lesson from a cookbook), there are four main groups of fats: monounsaturated, polyunsaturated, trans fats and saturated. Different foods sit within each group.

"GOOD" FATS

MONOUNSATURATED

Avocados

Olives

Nuts (almonds, peanuts, macadamia nuts, hazelnuts, pecans, cashews)

Nut butter (containing just nuts and salt)

POLYUNSATURATED FATS

Walnuts

Sunflower, sesame, and pumpkin seeds

Flaxseed

Fatty fish (salmon, tuna, mackerel, herring, trout, sardines)

Soy milk and tofu

"BAD" FATS

TRANS FATS

Commercially baked goods (e.g. cookies, crackers, cakes, muffins)

Packaged snack foods (e.g. crisps, sweets)

Solid fats (e.g. margarine)

Fried foods (e.g. chips, fried chicken, chicken nuggets)

SATURATED FATS

Processed meats like sausages and burgers

Hard cheeses

Fatty meat

Palm oil

TRANS FATS

Artificial trans fats can be found in many processed foods, and are generally speaking ones to avoid as much as possible.

SATURATED FATS

Don't be afraid of all saturated fat. The saturated fat in nuts is different to the saturated fat found in pizza, for instance. And just as saturated fat varies according to its source, the effect of saturated fats on blood cholesterol can vary from person to person, depending on genetics and other health factors.

Generally speaking, though, it's all about applying a bit of common sense to the way you eat. You don't need to cut out fats from your diet; instead try to replace some of the "bad" fats with a few more of the "good" ones, and you'll be flying.

A QUICK WORD ON PALM OIL

On all our packaging we say "Absolutely no palm oil" and we get asked all the time what the deal is with it.

Palm oil is taken from the fruit of the oil palm tree and originates in western Africa, but flourishes anywhere where heat and rainfall are abundant. The world's biggest producers and exporters of palm oil are Indonesia and Malaysia.

It can be used in quite literally everything and anything. If you check out the ingredients in products you'll start seeing it popping up in all sorts of places, from lipstick to instant noodles and peanut butter.

In the case of peanut butter, palm oil is used as an emulsifier – a fancy word for saying it keeps ingredients stuck together. Peanut butter with palm oil is much thicker than natural peanut butters without it, which will have a runnier texture and will often have oil sitting at the top.

The problem with palm oil is that the industry is linked to major issues such as deforestation, habitat degradation, climate change, animal cruelty and indigenous rights abuses in the countries where it is produced, as the land and forests must be cleared for the development of the oil palm plantations. Sadly, at least half of the world's orangutans have disappeared in the last 20 years, with over 80% of their habitat either depopulated or completely destroyed as a result of this deforestation.

Work is being done by organizations like the Roundtable on Sustainable Palm Oil, who work with plantations to ensure palm oil is sustainable and complies with various globally set standards.

The other downside is that palm oil has higher levels of saturated fats compared to other vegetable fats, and studies have shown a link between the intake of palm oil and high levels of cholesterol.

2. DIY NUT BUTTER

DIY NUT BUTTER

Nut butters are really easy to
make; in fact your food processor
does all the work for you. The trick
is to be patient and wait until the
nut butter is really smooth and glossy.
You'll be tempted to stop the blender after
6 minutes, but honestly it's worth the wait if
you leave it running for a few minutes more.

There are just a couple of words of warning before you start:
Making nut butter is pretty noisy, so perhaps wait until the
house has woken up before you crack on with making your own.

Making nut butter requires a pretty powerful food processor and
has been known on occasion to send a few off to food-processor
heaven. Be sure to keep an eye on your processor, as the friction
from the milling can sometimes cause the blender to overheat.

PLAIN NUT BUTTER

- 300g/2–2½ cups nuts
- Pinch of sea salt (optional)

MAKES
1 JAR

TAKES
10–12
MIN

Preheat the oven to 150°C/300°F/gas mark 2. Spread the nuts out on a large baking tray in a single layer. Roast until golden brown, about 10–15 minutes. Keep an eye on them, as nuts can turn from golden brown to burnt very quickly! Generally speaking, when you start smelling them around the house it's time to take them out – pronto!

Tip the nuts into a food processor with the salt, if using, and blitz. The whole process takes about 10 minutes, depending on how powerful your blender is, but don't stop until the nut butter is smooth and glossy:

After 2 minutes: The nuts turn into a crumble-like texture. At this point, stop the blender and use a spatula to scrape down all the pieces.

After 4 minutes: A small ball forms.

After 6 minutes: The nuts will be smooth, but will still have a slightly rough texture.

After 8–10 minutes: Don't stop yet. Keep going until the nut butter is smooth, glossy and runny.

Pour into an airtight container or jam jar. It will be good for 3 months, but to get the best flavour from the nuts, eat within 2 weeks or so.

BLEND ALL YOUR DRY INGREDIENTS THEN ADD THE WET ONES*

*we're making Coconut Chia Almond Butter, see page 30

BLEND TIL SMOOTH AMD GLOSSY

ACTIVATED NUT BUTTER

You might like to activate your nuts before making them into nut butter. Activation increases the nutrient value of the nuts and helps make them easier to digest. It's a bit of a fiddly process but if you have the time it's worth it.

Wash the nuts, then pour enough cold, fresh tap water into a pan or bowl to cover the nuts when they will be added. Add table salt to the water (for amounts, see guide below) and stir to dissolve, until you can't see any grains. Then place the nuts in the water and leave to soak (see guide below for soaking times).

Drain the nuts and rinse under cold running water. Spread out evenly on a baking tray in a single layer and dry the nuts on the lowest setting on your oven (for timings, see guide below), turning them occasionally.

Once transformed into nut butter, eat within 1–2 days.

NUT/SEED	QUANTITY	SOAK TIME	DRY TIME
Almonds	1 tsp salt per ½ cup nuts	7–12 hours	12–24 hours
Cashews	1 tsp salt per ½ cup nuts	3–6 hours	12–24 hours
Hazelnuts	1 tsp salt per ½ cup nuts	7–12 hours	12–24 hours
Peanuts	1 tsp salt per ½ cup nuts	7–12 hours	12–24 hours
Pine Nuts	1 tsp salt per ½ cup nuts	7–12 hours	12–24 hours
Pecans	1 tsp salt per ½ cup nuts	7–12 hours	12–14 hours
Walnuts	1 tsp salt per ½ cup nuts	7–12 hours	12–14 hours
Pumpkin Seeds	2 tsp salt per ½ cup seeds	7–12 hours	12–14 hours
Sunflower Seeds	2 tsp salt per ½ cup seeds	7–12 hours	12–14 hours

FLAVOURED NUT BUTTERS

MAKES
1 JAR

Playing with different nut butter flavours in our kitchen is what we do best. There are endless recipes but on the following pages is just a handful of our favourites.

1. MIXED NUT BUTTER

- 150g/1½ cups pecans
- 150g/scant 1¼ cups brazil nuts
- 150g/1¼ cups cashew nuts
- pinch of sea salt

Roast all the nuts in a single layer, following the method on page 27, then place in the food processor with the salt and blitz until smooth and glossy (see page 27).

Pour into an airtight container.

2. HONEY PISTACHIO BUTTER

- 200g/1½ cups shelled pistachios
- 1 Tbsp coconut oil
- pinch of sea salt
- 2 tsp runny honey

Put the pistachios into a food processor and blitz until smooth and glossy (see page 26). Add the coconut oil and salt, then blitz until fully melted and combined. Blitz in the honey 1 teaspoon at a time. If you find that it starts to dry up then just add some more coconut oil to loosen it.

Pour into an airtight container.

3. ESPRESSO ALMOND BUTTER

- 220g/scant 1¾ cups whole almonds
- pinch of sea salt
- 2 Tbsp freshly made espresso
- about 1 Tbsp agave nectar

Roast and grind the almonds following the method on page 27, and once the almond butter is smooth and glossy add the salt with the espresso, 1 tablespoon at a time, blitzing between additions. Blitz in the agave nectar gradually, as you may find you only need ½ tablespoon, depending on how finely you have milled your almonds.

Pour into an airtight container.

4. COCONUT CHIA ALMOND BUTTER

- 200g/1½ cups whole almonds
- 1 Tbsp chia seeds
- 75g/2½oz creamed coconut
- pinch of sea salt
- ½ Tbsp agave nectar

Roast the almonds following the method on page 27, then add to the food processor with the chia seeds and blitz until smooth and glossy (see page 26). Add the creamed coconut and salt and blitz until smooth. Add the agave nectar gradually and mix until well combined.

Pour into an airtight container.

MACADAMIA BUTTER

PEANUT BUTTER

HONEY PISTACHIO
BUTTER

COCONUT CHIA ALMOND BUTTER

ALMOND BUTTER

3

Breakfast & Brunch

SIX WAYS WITH TOAST

Toast was voted one of the UK's favourite smells (back in a very official-looking *Daily Mail* survey in 2010). It's up there with the smell of bacon, freshly cleaned houses and roast dinners. And for good reason, we say. There aren't many problems that a slice or two with a good cup of tea can't solve. Here are six recipes for you to create.

The bread maketh the toast. So choose wisely.
Go for an unsliced loaf rather than something pre-sliced. We've suggested a bread type in each of the recipes but generally speaking any of the below will work well with the recipes in this section.

- Rye (light or dark)
- Sourdough (brown or white)
- Multigrain or seeded loaf
- Spelt loaf

1. PECAN BUTTER, GRATED APPLE AND MAPLE TOAST

The pecan and apple on this nut butter give this slice a taste of the US.

MAKES
2 SLICES

TAKES
5 MIN

- 2 slices of rye bread
- 4 Tbsp pecan butter
- 1 apple, cored and grated
- Drizzle of maple syrup

Toast the rye bread and, while still warm, spread with the pecan butter. Top with the grated apple and drizzle with maple syrup.

2. AVOCADO, ALMOND BUTTER AND SPICED SUNFLOWER SEEDS TOAST

Avocados are awesome. Almonds are ace. Altogether they are amazing. Think that's enough alliteration for one recipe.

MAKES 2 SLICES

TAKES 5 MIN

- 2 Tbsp sunflower seeds
- ¼ tsp paprika
- ¼ tsp ground cumin
- Pinch of sea salt
- 2 slices of multigrain bread
- 4 Tbsp almond butter
- 1 ripe avocado

Toast the sunflower seeds in a dry pan over a medium heat for 2 minutes, then add the paprika, cumin and sea salt and toast for another 1 minute.

Meanwhile, toast the bread and spread with almond butter while still warm. Peel and thinly slice the avocado. Lay the avocado on the toast and sprinkle with the seeds.

3. MARINATED STRAWBERRIES AND ALMOND BUTTER SOURDOUGH

The savoury edge from the almond butter, sweet honey, salty sourdough and tart lemon makes this a real crowd-pleaser. Plus saying that you marinated your strawberries makes it sounds rather fancy, so you'll get lots of praise from friends if you're generous enough to make it for them. It's a win-win situation.

MAKES
2 SLICES

TAKES
10 MIN

- Juice of 1 lemon
- ½ vanilla bean pod, split in half lengthways and seeds scraped
- About 1 Tbsp honey
- 250g/9oz strawberries, hulled
- 2 large slices of sourdough bread, halved, or 4 regular slices (thickly sliced)
- 8 Tbsp almond butter
- 2 tsp pumpkin seeds (optional)

Put the lemon juice, vanilla seeds and honey in a bowl and mix well with a fork to combine. Taste and add more honey if it seems too tart.

Slice the strawberries lengthways into heart-shaped slices and add to the syrup mixture. Mix well until all the strawberries have a good coating.

Toast the sourdough slices until golden, then immediately spread each slice generously with almond butter. Add the strawberry mixture and top with a sprinkling of pumpkin seeds, if you like a little extra crunch.

4. FIG, GOAT'S CHEESE AND WALNUT BUTTER WITH HONEY TOAST

Think of this as a deconstructed version of the classic fig and goat's cheese salad, but served on top of a really crusty piece of toast. You can get all cheffy with this recipe by drizzling the honey from a great height. It'll look dead impressive.

MAKES 2 SLICES

TAKES 5 MIN

- 2 slices of walnut bread
- 4 Tbsp walnut butter
- 2 figs, quartered
- 100g/3½oz goat's cheese
- 2 Tbsp honey

Toast the bread slices, then spread immediately with the walnut butter, top with the fig quarters and crumble over the goat's cheese. For the final touch, drizzle the honey on top.

PIP'S TOP TIP:

Figs are in season late summer in the UK so if you struggle to get your hands on them, you can can replace with some slices of pear on top.

5. HAZELNUT BUTTER WITH YOGHURT AND DARK CHOCOLATE TOAST

Watch out for the cheeky holes in the sourdough which have a pesky way of allowing some of the topping to drip and inevitably land on you when eating. We've made that mistake one too many times.

MAKES 2 SLICES

TAKES 5 MIN

- 50g/⅓ cup hazelnuts
- 2 slices of sourdough bread
- 4 Tbsp hazelnut butter
- 3 Tbsp Greek yoghurt

TO SERVE
- Grating of dark chocolate

Toast the hazelnuts in a dry pan over a medium heat for 2–3 minutes. Keep an eye on them as they can turn quickly.

Meanwhile, toast the sourdough slices and spread generously with hazelnut butter while still warm. Dollop the Greek yoghurt on top, then swirl it through the hazelnut butter. Sprinkle some dark chocolate over the top of both slices. For a final touch, roughly chop the hazelnuts and sprinkle over before eating.

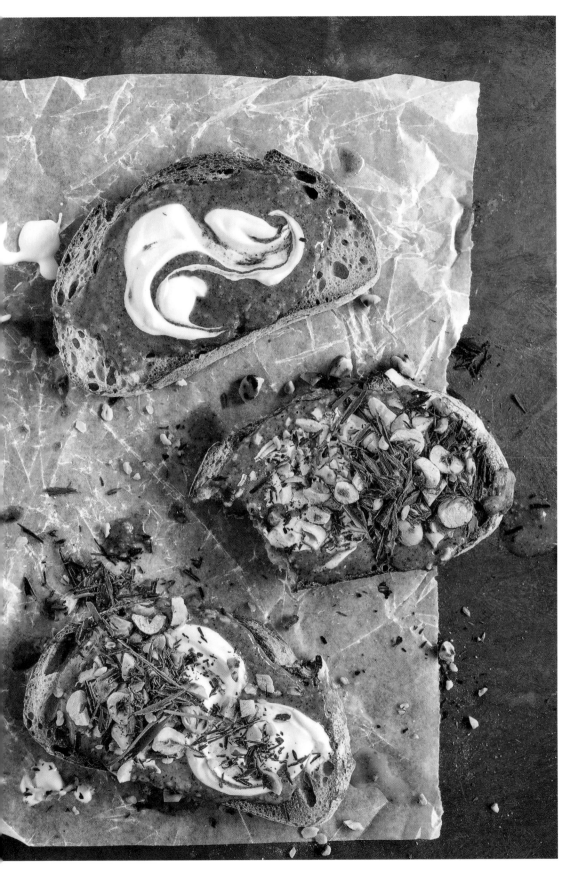

6. PEANUT BUTTER, CHILLI, LIME AND CORIANDER ON TOAST

Apparently when you eat chilli a burning sensation is sent from the nerve endings in the mouth to the brain. The body defends itself against this pain sensation by secreting endorphins – natural painkillers that cause a physical "rush" – a high that keeps us craving more. So, depending on how much of a chilli junkie you are, you might need to up the ante on this one and keep the seeds in the chilli for more heat.

MAKES
2 SLICES

TAKES
2 MIN

- 2 slices of brown seeded or multigrain loaf
- 4 Tbsp peanut butter
- ½ red chilli, deseeded and very finely chopped
- Handful of coriander (cilantro), leaves only
- ½ lime

Toast the bread slices then immediately spread with the peanut butter. Sprinkle with the chopped chilli and coriander (cilantro) leaves, and squeeze a quarter of lime over the top of each.

TOAST OF THE TOWN

ELVIS PRESLEY TOASTED SANDWICH

The Elvis Presley sandwich, otherwise know as the Fool's Gold Loaf, is now a well fabled story whereby Elvis would combine a pound of bacon, peanut butter and grape jelly inside a warm loaf of Italian bread. Now, we're not suggesting you go quite to the super-size lengths that these tales suggest, or in fact eat one of these every day, but he certainly got something right with the flavour combination. If it's good enough for the King, then it's good enough for us, we say.

- 4 slices of streaky bacon
- 4 slices of white thick bread
- 6 Tbsp peanut butter
- 2 bananas, sliced
- Drizzle of honey or maple syrup (optional)
- 3 Tbsp butter, softened

Preheat the grill to high and cook the bacon on each side for 2–3 minutes, depending on how crispy you like it. Transfer to a plate lined with kitchen paper.

Meanwhile, very lightly toast the bread for 30 seconds, then spread with the peanut butter. Place the banana slices on 2 of the slices and add 2 slices of bacon to each of the other 2 toast slices. Drizzle over some honey or maple syrup; then sandwich together and spread the butter on the outside of all the pieces of toast.

Heat a large frying pan over a medium heat and, when hot, add the sandwiches to the pan. Toast until golden brown on both sides, about 5 minutes. Cut in half to serve.

OVERNIGHT OATS THREE WAYS

Just reached the end of your jar? There's no need to despair – make overnight oats! These recipes work best if you use your almost-empty jars, as the oats absorb the last nut butter scraps inside.

And if you're having one of those mornings (you know, the ones where your alarm didn't go off, the kids are having a pre-school tantrum or your car won't start) you can just grab your jar and head out of the door. Pronto.

 PLUS 12 HOURS OF SOAKING

1. CLASSIC OVERNIGHT OATS

- 50g/⅓ cup oatmeal
- 160ml/⅔ cup almond milk
- 2 Tbsp chia seeds
- ½ apple, grated
- 1 Tbsp cashew butter
- 50g/1¾oz fruit (berries, chopped banana), plus extra to serve

Simply add all the ingredients to the jar, give it a good mix and leave in the fridge to absorb all the liquid overnight. When you get up in the morning just empty the jar into a bowl and add some more nut butter and fruit.

PIP'S TOP TIP:

If the jar isn't large enough or you haven't got one to hand just put the oat mixture in an airtight container.

2. CARROT CAKE OVERNIGHT OATS

- 60ml/¼ cup almond milk
- 50ml/scant ½ cup
  full-fat natural yoghurt
- ½ carrot, grated
- ½ tsp ground cinnamon
- 50g/⅓ cup oatmeal
- 1 tsp agave nectar
- 1 Tbsp nut butter, plus extra
  to mix through in the morning

Simply add all the ingredients to the jar, give it a good mix and leave in the fridge to absorb all the liquid overnight. When you get up in the morning just empty the jar into a bowl and add some more nut butter and fruit.

3. BIRCHER MUESLI

- 50g/½ cup oatmeal
- 1 apple, grated
- 50ml/scant ¼ cup
  apple juice
- 100ml/scant ½ cup
  full-fat natural yoghurt
- 1 medium banana, sliced
- 30g/3 Tbsp sultanas
  (golden raisins)
- ¼ tsp ground cinnamon
- 10g/⅓oz flaked
  (slivered) almonds
- 15g/½oz almond butter

Simply add all the ingredients to the jar, give it a good mix and leave in the fridge to absorb all the liquid overnight. When you get up in the morning just empty the jar into a bowl and add some more nut butter and fruit.

FOUR WAYS WITH PORRIDGE

These porridge recipes are pretty great. We think even Goldilocks would approve. We know that the purists will probably frown upon our rather unconventional technique, but by adding a little nut butter during the cooking stage you get a subtle nuttiness running throughout your porridge which is pretty darn tasty.

Pip's Top Tip on Porridge:

- Make sure you use oatmeal rather than rolled oats to get a better bite and avoid it turning into school-dinner-style gloop.
- Add a pinch of salt to the oats before cooking. It works. Trust me.
- Try soaking your oats in liquid overnight and they'll cook even more quickly in the morning.
- Or if you want to have a more oat-y flavour running through your bowl, try toasting the oats over a low heat in a dry pan for a couple minutes before cooking.

1. CACAO, HAZELNUT AND BANANA PORRIDGE

Don't get mixed up when shopping for cacao. Just a pesky switch of the a and o makes a very different porridge.

FOR THE OATS
- 75g/½ cup oatmeal
- 225ml/scant 1 cup hazelnut milk (or cow's milk)
- 2 tsp cacao nibs
- Pinch of salt
- 4 Tbsp hazelnut butter

FOR THE TOPPING
- 1 large banana, sliced
- 2 Tbsp hazelnuts, roughly chopped
- 2 Tbsp honey (optional)

Put the oatmeal and milk into a medium pan over a medium heat. Cook, stirring constantly with a wooden spoon, until bubbling, then turn down the heat and simmer, stirring, until creamy. Stir in the cacao nibs and salt, then take off the heat.

Briefly stir the hazelnut butter into the porridge and divide between 2 bowls. Add the sliced banana to the top, sprinkle with the chopped hazelnuts and drizzle with honey, if you want extra sweetness.

2. CASHEW, BLACKBERRY AND PEAR PORRIDGE

If you're making this recipe in August or September, then forget going to the supermarket and instead head to your local hedgerows where blackberries are ripe and ready for the picking. Just be sure to actually collect some for your breakfast rather than gobbling them all up there and then.

SERVES 2

TAKES 15 MIN

FOR THE BLACKBERRY JAM
- 50g/⅓ cup fresh or frozen blackberries
- Finely grated zest of ½ lime
- 2 tsp honey
- 3 Tbsp water

FOR THE OATS
- 75g/½ cup oatmeal
- ½ tsp ground cinnamon
- 50ml/scant ¼ cup milk
- 200ml/¾ cup water
- 1 Tbsp cashew butter
- Pinch of sea salt

TO SERVE
- 3 Tbsp cashew butter
- 1 pear, cored and thinly sliced

For the blackberry jam, put the blackberries and lime zest, honey and water into a pan over a medium-low heat and bring to a simmer. Continue to cook for another 5 minutes, stirring occasionally, but being careful not to mush the blackberries. Take off the heat once you get a jam-like consistency.

Put the oatmeal, cinnamon, milk, water and cashew butter in a separate, medium pan and bring to a simmer over a medium heat, stirring frequently with a wooden spoon. When the oats start to bubble, turn down the heat and cook, stirring, for another 5–7 minutes to achieve a nice creamy texture. Stir in the salt.

Divide the cooked porridge between 2 bowls and ripple the cashew butter through each bowl. Top with the blackberry jam and pear slices and serve.

3. COCONUT AND ALMOND BUTTER PORRIDGE

Like a macaroon but in a hot steamy bowl of porridge.

FOR THE OATS

- 1 Tbsp creamed coconut
- 1 Tbsp almond butter
- 75g/½ cup oatmeal
- 225ml/scant 1 cup water
- Pinch of sea salt

FOR THE TOPPING

- 2 tsp flaked (slivered) almonds
- 4 Tbsp coconut yoghurt
- 2 Tbsp almond butter
- 2 tsp coconut sugar
- 75g/scant ⅔ cup blueberries

Toast the flaked (slivered) almonds in a dry pan over a medium heat until golden. Remove from the heat and tip onto a plate.

Put the creamed coconut, almond butter, oatmeal and water into a medium pan over a medium heat, stirring continuously using a wooden spoon. Once you see the porridge bubbling, turn down the heat and cook, stirring, for another 5-7 minutes, until the oats are soft and creamy.

Divide the porridge between 2 bowls and top with the coconut yoghurt and almond butter, then a sprinkling of coconut sugar, toasted flaked almonds and blueberries.

4. ALMOND, RAISIN AND APPLE PORRIDGE

Raisins have a lovely way of softening up in this recipe, but if you like them all plumped up then just soak overnight in a little water and they will be even more juicy.

FOR THE OATS
- 75g/½ cup oatmeal
- 100ml/scant ½ cup milk
- 150ml/⅔ cup water
- 50g/⅓ cup raisins
- ½ tsp ground cinnamon
- 2 Tbsp almond butter

FOR THE TOPPING
- 2 Tbsp almond butter
- 1 apple, grated (skin on)
- 4 Tbsp maple syrup

Put the oatmeal into a medium pan with the milk, water, raisins, cinnamon and almond butter. Bring to the boil over a medium heat, stirring continuously, then turn down the heat and cook, stirring, until the porridge is creamy and the raisins have softened, for about 5–7 minutes.

Divide the porridge between 2 bowls then swirl in the almond butter, pile the grated apple on top and add a generous drizzle of maple syrup.

PEP UP YOUR PORRIDGE

COCO-NUTTY BANANA PANCAKES

Flippin' amazing pancakes. You'll go absolutely bananas for these guys. Or nuts. Whatever cheesy pun you want to use to celebrate just how tasty these light, fluffy pancakes are.

- 2 ripe bananas, cut into chunks
- 1 egg
- 100ml/scant ½ cup almond milk
- 2 Tbsp almond butter
- 1 Tbsp coconut nectar
- 1 tsp vanilla extract
- 110g/¾ cup plus 1½ Tbsp plain (all-purpose) flour
- 50g/6 Tbsp coconut flour
- Pinch of ground cinnamon
- 1 tsp baking powder
- Pinch of bicarbonate of soda (baking soda)
- Pinch of salt
- Coconut oil, for frying

FOR THE TOPPING
- Almond butter
- Fresh berries, such as strawberries, blueberries, blackberries
- Coconut yoghurt
- Honey

In a food processor, blend the bananas with the egg, almond milk, almond butter, coconut nectar and vanilla, until smooth and well combined.

Sift together the dry ingredients in a large bowl then slowly add the wet ingredients to the dry ingredients, whisking together until you have a thick, smooth batter.

Working in batches, add 1 tablespoon coconut oil to a large, non-stick frying pan and heat until melting. Once the pan is nice and hot, drop large tablespoonfuls of the batter into the pan and cook for 2 minutes on each side until lightly browned and the pancakes are fluffy. If you're making a big batch to serve all at once, then keep the cooked pancakes on a heatproof dish, separated with baking parchment, in a low oven, while you cook the rest.

To serve, top the pancakes with almond butter, berries, yoghurt and honey.

BAKED PECAN BRIOCHE FRENCH TOAST

In France they actually call French toast "pain perdu" (lost bread), and make it using stale, leftover bread. Well, we thought that sounded a bit sad so we've opted for brioche to help give this breakfast an added "je ne sais quoi".

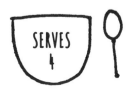

SERVES
4

TAKES
35 MIN

PLUS 12 HOURS OF SOAKING

- 1 loaf of brioche, sliced fairly thickly
- 200g/scant 1 cup pecan butter, plus extra to serve
- 4 eggs
- 55g/scant ⅓ cup caster (superfine) sugar
- 500ml/2 cups almond milk
- 100g/generous ¾ cup fresh blueberries, plus extra to serve
- Icing (confectioners') sugar, for dusting

TO SERVE
- Maple syrup
- Natural yoghurt

The day before, spread one side of each brioche slice with pecan butter and arrange the slices in overlapping layers in a shallow ovenproof dish, with the nut butter sides facing down.

In a large bowl, whisk the eggs, sugar and almond milk until combined, then pour evenly over the brioche. Dot the blueberries all over the surface, cover the dish with cling film and refrigerate overnight.

The next day, preheat the oven to 175°C/350°F/ gas mark 4. Remove the cling film from the dish, place in the oven and bake for 30 minutes, or until golden brown and puffed up, and no more liquid remains. Leave to cool a little to warm, then dust with icing (confectioners') sugar.

Serve with a maple syrup, a spoonful of yoghurt, some more blueberries and pecan butter.

BLUEBERRY AND ALMOND BREAKFAST MUFFINS

Make a batch of these muffins on a Sunday afternoon and your future self will be thanking you all week as you happily unwrap your muffin and enjoy with a nice cup of tea.

- 80g/⅓ cup almond butter
- 200g/7oz Greek yoghurt
- 1 large ripe banana, cut into chunks
- 1 large egg
- 1 tsp vanilla extract
- 120g/scant ⅔ cup light brown sugar
- 220g/1⅔ cups plain (all-purpose) flour
- 50g/½ cup rolled oats
- 1 tsp baking powder
- ½ tsp bicarbonate of soda (baking soda)
- ½ tsp salt
- 65g/½ cup blueberries
- 75g/2½oz medjool dates, pitted and chopped

FOR THE TOPPING
- 2 Tbsp light brown sugar
- 50g/⅔ cup flaked (slivered) almonds
- 2 Tbsp seeds, such as sesame, poppy, pumpkin or sunflower

Preheat the oven to 160°C/325°F/gas mark 3. Line a muffin tray (or trays) with muffin cases or squares of baking parchment.

Put the almond butter, yoghurt, banana, egg, vanilla and sugar into the bowl of a stand mixer and beat well until the mixture forms a batter. Stir in the flour, oats, baking powder, bicarbonate of soda and salt until just combined. Stir in the blueberries and dates, using a wooden spoon.

Spoon the batter into the muffin cases to about three-quarters full. Sprinkle each muffin with the light brown sugar, flaked (slivered) almonds and seeds.

Bake for 20–25 minutes until cooked, then leave to cool in the tray/s on a wire rack. Store in an airtight container.

"MORNIN' SUNSHINE" GRANOLA

Get this wrong and we'll eat our hats. Well, maybe not our hats, but we'll be shocked, for sure! Generally speaking the more you chuck into the mix the tastier your granola's going to get, and the better your morning's going to be.

**MAKES
1 JAR**

**TAKES
55 MIN**

- 3 heaped Tbsp nut butter (almond is our favourite)
- 3 Tbsp coconut oil
- 4–5 Tbsp honey or maple syrup
- ½ tsp ground cinnamon
- 300g/3 cups rolled oats
- 100g/scant ⅔ cup quinoa, rinsed (if you like it crunchy)
- 50g/1¾oz brown puffed rice (optional)
- 200g/1½ cups mixed nuts, roughly chopped (try a blend of almonds, pistachios, pecans)
- 50g/5½ Tbsp pumpkin seeds and/or sunflower seeds
- 50g/1 cup coconut shavings
- 1 large egg white, whisked (optional)
- 80–100g/½–¾ cup raisins, chopped dried apricots, dates or cherries

Preheat the oven to 150°C/300°F/gas mark 2.

Combine the nut butter, coconut oil, honey or maple syrup and cinnamon in a saucepan and melt over a low heat. Mix the remaining ingredients, except the egg white and dried fruit, in a bowl, then stir in the melted mixture to combine. Add the whisked egg white at this stage if you want bigger clusters, as it help everything stick better (but is just as good without).

Spread the mixture out in an even layer on 2 large baking trays and bake for 35–45 minutes, or until it looks nicely toasted, keeping an eye on it so it doesn't over-bake, and stir occasionally if you don't want it in clumps. Once completely cooled, stir in the chopped fruit and store in a jar.

PIP'S TOP TIP:

You can reduce the amount of oats and add the equivalent amount of another grain, such as spelt, barley or rye flakes.

BAKEWELL TART WAFFLES

The combination of cherry and almond in these waffles was inspired by the good ole' classic English pud, the Bakewell tart. A Sunday lunch classic that we've squashed in between two waffle irons.

- 100g/¾ cup plain (all-purpose) flour
- 50g/½ cup almond flour
- 1 tsp baking powder
- 1 tsp ground cinnamon
- 2 eggs
- 225ml/scant 1 cup almond milk
- ½ Tbsp coconut oil, melted
- 120g/½ cup almond butter
- 150g/5¼oz cherries, pitted and quartered

TO FINISH AND SERVE
- Icing (confectioners') sugar
- Coconut yoghurt
- Almond butter
- Cherries, pitted

Heat up your waffle iron.

Sift both flours, the baking powder and cinnamon together into a bowl. Mix the eggs and almond milk together in a separate bowl and gradually add to the flour mixture to make a smooth batter. Stir the melted coconut oil into the mixture, then stir through the almond butter and cherries.

Once the waffle iron is very hot, ladle in the batter and cook until nice and crispy on the outside and they come away easily from the sides (follow the manufacturer's instructions). Dust with icing (confectioners') sugar, top with coconut yoghurt, drizzle with almond butter and scatter with more cherries.

4. SNA

THE GO!

_____

_____

_____

PICK YOUR OWN PROTEIN BALLS

This recipe is as easy as A-B-C or 1-2-3, as the Jackson 5 would say. Just follow the steps below and you'll be snack-time-ready in next to no time.

1 THE BASE

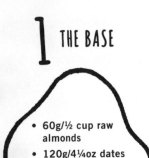

- 60g/½ cup raw almonds
- 120g/4¼oz dates
- 40g/⅓ cup dried cherries

2 PICK YOUR NUT BUTTER (2 TSP)

- Almond butter
- Peanut butter
- Cashew butter
- Hazelnut butter
- Pecan butter
- Coconut almond butter
- Seed butter

MAKES 12-15

TAKES 15 MIN

If you're not sure where to start with this one, then these balls have been tried, tested and devoured:

- Almond butter, pecan nut, lucuma ball
- Cashew butter,
- Coconut, maca ball
- Peanut butter,
- Cacao ball
- Peanut butter, apple and cinnamon ball

Blitz the base ingredients in a food processor for 3–5 minutes, or until it forms a sticky dough.

Pick your nut butter of choice and choose your filler (we recommend a max of two fillers per mixture). Add both to the food processor and blitz until combined but not completely smooth; you want to keep a bit of texture.

Take a small amount of mixture and roll it between your hands to make a small, golf-ball sized bite. Pick your topping and spread it onto a large plate, then roll each ball around until nicely coated. Repeat with the remaining mixture to make more balls.

Store refrigerated in an airtight container, for up to a week.

3 PICK YOUR FILLER (TWO MAX.)

- 20g/½ cup coconut flakes
- ½ Tbsp maca powder
- ½ Tbsp lucuma powder
- 20g/3 Tbsp chopped nuts (pecans, macadamia, pistachio)
- 1 Tbsp chia seeds
- 1 Tbsp ground cinnamon
- 1 Tbsp cacao powder
- 20g/3 Tbsp chopped dried apple

4 PICK YOUR TOPPING (ONE MAX.)

- Sesame seeds
- Cocoa powder
- Finely chopped pistachios
- Finely chopped mixed nuts
- Desiccated coconut
- Cacao nibs

CHOCOLATE-DIPPED NUT BUTTER CRISPY BITES

The adult version of the childhood Rice Krispies cakes. We recommend sticking to the recipe below – they're pretty tasty – but if the urge takes you, feel free to add some mini chocolate eggs to the top for added nostalgia.

MAKES 16

TAKES 3 HRS

A little vegetable oil, for greasing

FOR THE BASE
- 3 Tbsp butter
- 6 Tbsp maple syrup
- 70g/2½oz crisped rice cereal

FOR THE CHOCOLATE PEANUT BUTTER LAYER
- 85g/3oz dark chocolate, broken into pieces
- 200g/scant 1 cup peanut butter

FOR THE CHOCOLATE FROSTING
- 85g/3oz dark chocolate (60–72% cocoa solids), roughly chopped
- ½ tsp agave nectar
- 4 Tbsp coconut oil

Line a 20cm/8-inch square baking tin and lightly oil.

Melt the butter and maple syrup together in a pan, then pour over the crisped rice cereal in a bowl. Working quickly, stir until the cereal is thoroughly coated, then pour it into the prepared tin. Using your hands, press the mixture into the base of the tin. Leave to cool to room temperature.

For the chocolate peanut butter layer, put the chocolate and peanut butter into a large heatproof bowl and set the bowl over a saucepan of simmering water, making sure the base of the bowl is not touching the water. Stir until melted and the mixture is smooth. Remove the bowl from the pan and leave to cool for 5 minutes, then pour the mixture over the base. Refrigerate for 1 hour, or until the top layer hardens.

For the chocolate frosting, combine the chocolate, agave nectar and coconut oil in a large heatproof bowl and melt as for the peanut butter layer, until completely smooth. Remove the bowl from the pan and cool for 5 minutes, then pour over the chocolate peanut butter layer and spread it out evenly. Refrigerate for 1 hour, or until the topping hardens.

Cut into 16 squares and serve, or store in the refrigerator in an airtight container for up to 4 days.

CASHEW OATMEAL DROPS

Pop these little biscuits into your lunchbox and you'll have the perfect snack for elevenses, or any time you get a bit peckish during the day.

MAKES
20

TAKES
30 MIN

- 280g/scant 3 cups rolled oats
- 180g/1½ cups cashews, chopped into small pieces
- 200g/scant 1 cup cashew nut butter
- 320g/11½oz agave nectar
- 240g/1¾ cups raisins
- 2 eggs, beaten

Preheat the oven to 180°C/350°F/gas mark 4. Line a couple of baking trays with baking parchment.

Combine all the ingredients in a bowl to make quite a wet but firm batter.

Drop tablespoonfuls of the batter onto the prepared trays, leaving plenty of space between each. Bake in the oven for 18 minutes, or until golden brown.

Cool on a wire rack and store in an airtight container for up to a week.

TEA TIME
TREAT

PEANUT BUTTER POPCORN CLUSTERS

Peanut-butter-covered popcorn in bite-sized clusters. You literally can't get more moreish than that.

MAKES
20

TAKES
15 MIN

- A little vegetable oil, for greasing
- 75g/⅓ cup butter
- 220g/1 cup plus 1 Tbsp brown sugar
- 2 Tbsp white granulated sugar
- 100ml/scant ½ cup double (heavy) cream
- Pinch of sea salt
- 100g/7 Tbsp peanut butter
- 50g/⅓ cup peanuts, chopped

FOR THE POPCORN
- 1 Tbsp butter
- 80g/scant ½ cup popcorn kernels

Lightly oil the insides of 20 muffin or cupcake tins.

Melt the butter for the popcorn in a large frying pan and pop the corn following the method on page 75. Tip into a bowl.

Put the butter, both sugars, cream and salt into a medium saucepan. Place over a medium-low heat and cook, whisking constantly, for about 4 minutes. Stir in the peanut butter and continue to whisk for another 3–4 minutes.

Remove from the heat, pour the caramel over the popcorn and stir to coat fully. Spoon the mixture into the oiled muffin or cupcake tins, sprinkle with the chopped peanuts and refrigerate until set.

SEA SALT CARAMEL POPCORN

The perfect accompaniment to any film marathon session, these are so good that we think they deserve an Oscar.

- 1 Tbsp butter
- 80g/scant ½ cup popcorn kernels
- 1 Tbsp coconut oil
- 2 Tbsp almond butter
- 3 Tbsp maple syrup
- 1½ tsp vanilla extract
- ½ tsp sea salt

Heat the butter in a large frying pan over a medium-high heat and, when nice and hot, add the popcorn kernels. Cover with a lid and, after about 30 seconds, you should start to hear the corn popping. Shake the pan a little to make sure all the kernels get popped. Once they sound as though they have stopped popping, remove and set aside in a bowl.

Put the coconut oil, almond butter, maple syrup and vanilla extract in a heavy-based saucepan. Stir together over a medium heat until the ingredients combine into a silky caramel, then add the sea salt.

Tip the caramel over the popcorn and, using a wooden spoon, coat the popcorn in the caramel.

FRUIT, VEG AND NUT BUTTER

Sometimes the simplest things in life are the best.
That's why we thought we'd show you a few basic food
combinations to pair with nut butter. No real cooking
involved here, more an assemble-nicely-on-a-plate
kind of dish.

TRY ANY OF OUR NUT BUTTERS COMBINED WITH:

MEDJOOL DATES

Stuff pitted dates with nut butter. One or two of
these bites will sort any sweet craving out.

BAKED FRUIT

Try baking your favourite pitted fruit in the
oven then top with yoghurt and nut butter
for a simple summer pud. See page 78 for our
apricot version.

CELERY & RAISINS

Otherwise known as ants on
a log. Fill the dip in the celery
with peanut butter and dot
with raisins.

BANANA
Slice a banana in half from end to end and spread with nut butter. It's like a dessert in its own right.

CARROTS & PEANUT
Dip carrot sticks into peanut butter. Sweet and nutty.

APPLE & PEAR SLICES
A classic combo. Simply core and slice apples and/or pears and dip/spoon with any nut butter you can get your hands on.

BAKED APRICOTS WITH ALMOND BUTTER

If it's summer and you're having a barbecue, try wrapping your apricots in foil and putting them on the grill while the coals are still warm. This makes for a lovely al fresco dessert, although if it is a truly British barbecue it will most likely be eaten sitting under a parasol in the rain. Apricot and almonds pair beautifully, but this dish also works very well with plums and nectarines.

- 6 apricots, halved with stones removed
- A little sunflower oil, for greasing
- 3 Tbsp almond butter
- Drizzle of honey
- 50g/⅔ cup flaked (slivered) almonds
- 6 Tbsp natural yoghurt

Wait until the BBQ coals are hot but no longer have any flames. Wrap the apricots loosely in foil and place on the grill for about 20 minutes.

You can also cook these in the oven: just preheat the oven to 200°C/400°F/gas mark 6 and lightly grease a baking tray.

Place the halved apricots on the baking tray and bake for 20 minutes, until nice and soft.

Drizzle the cooked apricots with the almond butter, some honey, the flaked (slivered) almonds and a few spoonfuls of yoghurt.

BBQ SPECIAL

SEEDED NUTTY ENERGY BARS

Not the prettiest little bar in the book, but never judge a snack by its looks. It's what's inside that counts.

- 2 Tbsp unsalted butter or coconut oil, plus a little extra for greasing
- 6 large medjool dates, pitted and chopped
- 2 Tbsp pumpkin seed butter
- 240ml/1 cup maple syrup
- 180g/scant 2 cups rolled oats
- 75g/½ cup almonds, hazelnuts, pecans, walnuts or cashews
- 50g/1¾oz pumpkin seeds
- 50g/1¾oz sunflower seeds
- ½ tsp salt

Preheat the oven to 180°C/350°F/gas mark 4. Lightly grease a 20 x 10cm/8 x 4-inch baking tray and line with baking parchment, leaving an overhang on the long sides.

Bring the dates, seed butter and maple syrup to the boil in a small saucepan, then reduce the heat to medium-high and boil, stirring often, until the dates are very soft and the maple syrup is slightly reduced, about 8–10 minutes. Remove from the heat and stir in the butter or coconut oil until melted. Mash the dates with a fork until as smooth as possible.

Toss the oats, nuts, pumpkin and sunflower seeds and salt together in a large bowl. Stir in the date mixture until evenly coated. Scrape half the oat mixture into the prepared tray and press very firmly and evenly with a rubber spatula to compress it as much as possible. Add the remaining oat mixture and press until it is very tightly packed into the tray.

Bake for 45–50 minutes, until the mixture has darkened in colour and is firm around the edges; the centre should give just slightly when pressed. Leave to cool in the tray before turning out (or leave to set overnight), then cut into 16 bars.

NUT BUTTER CUPS

Hold these nut butter cups under your chin and if your neck lights up brown then it's empirical proof that you like nut butter. No, wait. That's the flower. Maybe just stick to eating them.

MAKES
18

TAKES
1 HR

1. DARK CHOCOLATE ALMOND BUTTER CUPS WITH TOASTED QUINOA

FOR THE FILLING
- 4 Tbsp quinoa
- 200g/scant 1 cup almond butter

FOR THE CHOCOLATE CASES
- 325g/11½oz dark chocolate (at least 80% cocoa solids), broken into pieces
- 1 Tbsp coconut oil

First rinse the quinoa and tip into a large non-stick frying pan. Turn the heat to medium and dry out the grains, stirring to move them about the pan. Once all the liquid has evaporated, continue cooking for 10–15 minutes, until the quinoa has turned a nutty brown and starts to "pop". Remove from the heat.

Melt the chocolate and coconut oil for the cases in a heatproof bowl set over a pan of simmering water, making sure the base of the bowl is not touching the water. Use half the melted chocolate to fill mini paper cups or cupcake papers up to a third full. Refrigerate until hard and set the remaining melted chocolate aside.

In a separate bowl, mix together the almond butter with 3 tablespoons of the toasted quinoa. Put a blob into each paper case over the set chocolate, then cover with the remaining melted chocolate (warming it to melt if it has set). Sprinkle the remaining quinoa over the top and refrigerate again to set.

Keep the cups refrigerated in an airtight container for up to 1 week.

2. DARK CHOCOLATE HAZELNUT CUPS WITH GOJI BERRIES

FOR THE CHOCOLATE CASES
- 325g/11½oz dark chocolate (at least 80% cocoa solids), broken into pieces
- 1 Tbsp coconut oil

FOR THE FILLING
- 4 Tbsp goji berries, chopped into smallish pieces
- 200g/scant 1 cup hazelnut butter

Make the chocolate cases following the method on page 80, using half the melted chocolate. Refrigerate until hard.

Stir three-quarters of the chopped goji berries into the hazelnut butter and put a blob into each paper case. Cover with the remaining melted chocolate (warming it to melt again if necessary), sprinkle the remaining goji berries on top and refrigerate again to set.

Keep the cups refrigerated in an airtight container for up to 1 week.

3. MILK CHOCOLATE MAPLE PEANUT BUTTER CUPS WITH SEA SALT

FOR THE CHOCOLATE CASES
- 350g/12¼oz milk chocolate, broken into pieces

FOR THE FILLING
- 1 Tbsp coconut oil
- 200g/scant 1 cup peanut butter
- 1 tsp sea salt, plus extra for sprinkling
- 1 Tbsp maple syrup

Melt the chocolate in a heatproof bowl set over a pan of simmering water, making sure the base of the bowl is not touching the water. Use a third of the melted chocolate to fill mini paper cups or cupcake papers up to a third full. Refrigerate until hard and set the remaining melted chocolate aside.

Melt the coconut oil in a pan over a low heat then add to the peanut butter in a bowl, with the sea salt. Mix until well combined, then slowly stir in the maple syrup.

Put a small amount of the nut butter mixture into each set chocolate case and carefully pour over the remaining melted chocolate. Sprinkle with sea salt and refrigerate again to set.

Keep the cups refrigerated in an airtight container for up to 1 week.

4. WHITE CHOCOLATE PISTACHIO NUT BUTTER CUPS

FOR THE CHOCOLATE CASES
- 400g/14oz white chocolate broken into pieces

FOR THE FILLING
- 170g/¾ cup pistachio butter
- 3 Tbsp pistachios, chopped

Melt the white chocolate following the method on page 80. Use a third of the melted chocolate to fill the cupcake cases and put in the fridge to set. Set the remaining melted chocolate aside.

Put a small amount of the pistachio butter in each set chocolate case, then carefully pour over the remaining melted chocolate. Sprinkle the chopped pistachios on top.

Keep the cups refrigerated in an airtight container for up to 1 week.

Clockwise from top left:
Dark chocolate hazelnut cups, Milk chocolate maple peanut butter cups, White chocolate pistachio nut butter cups and Dark chocolate almond butter cups

Savoury

PEANUT SOBA NOODLE SALAD

Get your chopsticks out and have fun trying to pick up those pesky edamame beans.

- 225g/8oz buckwheat soba noodles
- 150g/5¼oz frozen edamame beans
- 2 Tbsp peanuts
- 2 carrots
- 3 spring onions (scallions), finely sliced
- 1 red chilli, finely diced
- 3 Tbsp coriander (cilantro) leaves, roughly chopped
- 3 Tbsp fresh mint leaves, roughly chopped
- 3 Tbsp mixture of black and white sesame seeds
- 1 lime, cut into wedges

FOR THE DRESSING
- 2cm/¾-inch piece of fresh ginger, peeled and grated
- 2 tsp sesame oil
- 2 tsp dark soy sauce
- 3 Tbsp rice vinegar
- 2 Tbsp rapeseed (canola) oil
- 2 Tbsp peanut butter
- 1 tsp coconut nectar

First make the dressing. Put all the ingredients into a small bowl, mix to combine and set aside.

Cook the noodles for 5–8 minutes in plenty of salted boiling water, stirring occasionally, until just tender. Drain and refresh by rinsing under cold water, then leave to dry on a clean tea towel.

At the same time, cook the edamame beans for 4–5 minutes in a separate pan of boiling water, then drain and refresh with cold water.

Roast the peanuts in a dry pan until golden, then take off the heat, tip onto a board and roughly chop. Peel the carrots and slice into fine matchsticks.

Put the noodles in a large mixing bowl and add the edamame beans, carrots, spring onions (scallions), chilli and herbs and toss with the dressing. Serve sprinkled with the chopped roasted peanuts, the sesame seeds and a wedge of lime.

SEARED TENDERSTEM BROCCOLI WITH GOAT'S CHEESE AND WALNUT SALAD

Tenderstem broccoli is just a taller version of a broccoli. Now, you could use the regular stuff, but if you want to be a bit posh then use this variety.

- 200g/7oz tenderstem broccoli
- 4 Tbsp extra virgin olive oil
- 2 tsp lemon juice
- ½ tsp honey
- 1 Tbsp walnut butter
- 1 tsp very finely chopped shallot
- 100g/3½oz goat's cheese
- Sea salt and freshly ground black pepper

Preheat the oven to 180°C/350°F/gas mark 4.

Put the broccoli with 2 tablespoons of the olive oil and some salt and pepper on a large baking sheet and toss to coat. Roast in the oven for about 25 minutes, turning it halfway through, until browned and tender and the leaves have gone a little crispy.

In a separate bowl, whisk the lemon juice with the honey, the remaining 2 tablespoons olive oil, the walnut butter and the shallot, and season with salt and pepper.

Place the broccoli in a bowl, drizzle with the walnut dressing and crumble over the goat's cheese.

CHEAT'S NUT BUTTER PESTO

We'll endorse a spot of cheating for the sake of making
sure you eat a good, healthy dinner.

1. ALMOND PESTO

- 50g/1¾oz Parmesan, grated
- 1 garlic clove, roughly chopped
- 80g/2¾oz basil leaves
- 3 Tbsp almond butter
- 75ml/5 Tbsp olive oil
- Sea salt and freshly ground black pepper

Blitz the Parmesan, garlic and basil with some
salt and pepper to a paste in a food processor.
Add the nut butter and pulse until combined.
Gradually blitz in the olive oil in a steady stream
until the mixture turns into a rough paste. Store
in an airtight container.

2. HAZELNUT PESTO

- 50g/1¾oz Parmesan, grated
- ½ garlic clove
- 80g/2¾oz flat-leaf parsley
- 3 Tbsp crunchy hazelnut butter
- Grated zest and juice of 1 lemon
- 75ml/5 Tbsp olive oil
- Sea salt and freshly ground black pepper

Follow the method for almond pesto, above, but
instead use hazelnut butter and parsley in place
of basil and adding the lemon zest and juice with
the nut butter.

3. SUNFLOWER SEED PESTO

- 50g/1¾oz Parmesan, grated
- 1 garlic clove, roughly chopped
- 50g/1¾oz coriander (cilantro), roughly chopped
- 3 Tbsp sunflower seed butter
- 75ml/5 Tbsp olive oil
- Sea salt and freshly ground black pepper

Follow the method for almond pesto, page 90, using coriander (cilantro) in place of basil, and the seed butter in place of nut butter.

PIP'S TOP TIP:

Try out these cheat's pesto recipes on your fave pasta, like this Green Pasta dish below.

GREEN PASTA

A zingy, zesty, nutty pasta dinner.

SERVES 4

TAKES 15 MIN

- 350g/12oz pasta (tagliatelle, pappardelle or spaghetti)
- 200g/7oz green beans, topped and tailed
- 4 Tbsp nut butter pesto (see page 90)
- Finely grated zest of 1 lemon
- 20g/¾oz parsley, roughly chopped
- 60g/2oz basil, roughly chopped
- Grated Parmesan
- Sea salt and freshly ground black pepper

Cook the pasta for 10–12 minutes in a pan of salted, boiling water until al dente, then drain, reserving 2 tablespoons of the cooking water in the pan.

Meanwhile, in a separate pan, boil the green beans for 4–6 minutes, then drain.

Return the just-drained pasta to the pan and add the drained beans and nut butter pesto. Stir to mix, then top with the lemon zest, herbs and a handful of grated Parmesan. Season to taste and serve.

PESTO CRUSTED FISH

Our fellow fish friends are having a bit of a hard time at the moment, so be sure to look out for the sustainable kind in your supermarket.

- 2 Tbsp nut butter pesto (see page 90)
- 85g/3oz fresh breadcrumbs
- Finely grated zest of 1 lemon
- 4 white fish fillets

Preheat the oven to 200°C/400°F/gas mark 6. Mix the pesto with the breadcrumbs and lemon zest.

Lay the fish skin-side down on a baking tray and press the breadcrumb and pesto mixture over each fillet, to cover the surface of the fish. Bake for 10 minutes, until just cooked through, then serve with a simple salad and some new potatoes.

NUTTY COUSCOUS SALAD

Couscous is typically served to celebrate a house warming in North Africa, so perhaps try bringing round a bowl next time you have a neighbour move in. You may get a few weird looks, but at least it's original.

- 150g/scant 1 cup couscous
- 1 stock cube
- 2 Tbsp olive oil
- 1 red onion, halved and finely sliced
- 5 Tbsp nut butter pesto (see page 90)
- 3 spring onions (scallions), finely sliced
- 50g/1¾oz rocket (arugula)
- Chopped nuts (to match those used in the pesto), for sprinkling
- Sea salt

Place the couscous in a bowl, crumble in the stock cube and pour boiling water over to come 5mm/⅕ inch above the couscous. Add ½ tablespoon of the olive oil, stir once, then cover with cling film.

Meanwhile, fry the red onion in the remaining oil over a low heat until completely soft. Season with salt.

When the couscous has softened, stir through the nut butter pesto. Add the red onion, spring onions (scallions) and rocket (arugula) and mix. Sprinkle over a handful of chopped nuts to serve.

PIP'S TOP TIP:

Keep a stock of jam jars at home. They are great for storing your pesto in or using them for shaking your salad dressings.

ROASTED VEG WITH NUT BUTTER PESTO

We love roasted veg as they are, but drizzling over your favourite nut butter pesto when they are hot out of the oven takes them to another level.

- 2 sweet potatoes
- 2 carrots
- 2 red onions
- 2 red peppers (bell peppers)
- Olive oil
- 3–4 Tbsp nut butter pesto (see page 90)
- Sea salt and freshly ground black pepper

Preheat the oven to 200°C/400°F/gas mark 6.

Peel the veg and chop into medium chunks, with the sweet potato in slightly smaller pieces than the other veg, and put into a roasting tin. Drizzle with olive oil, sprinkle with salt and pepper and use your hands to coat the veg all over.

Cook in the oven for about 1 hour, until nicely brown and caramelized, turning them halfway through, then drizzle with the nut butter pesto and serve.

CHICKEN SOUP WITH LEAFY GREENS AND SWEET POTATO

Chicken soup that really is good for the soul.

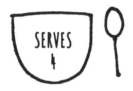

- 1 litre/4 cups chicken stock
- 1 large onion, thinly sliced
- 1 large sweet potato, peeled and chopped into 1.5cm/½-inch cubes
- 3 boneless, skinless chicken breasts
- 2 Tbsp finely chopped fresh ginger
- 100g/scant ½ cup almond butter
- 2 handfuls of spring greens, roughly chopped
- Sea salt and freshly ground black pepper

TO SERVE
- 1 lime, cut into wedges
- Chopped coriander (cilantro)

Put the stock, onion, sweet potato, chicken and ginger into a large pan and bring to the boil. Turn down the heat and simmer for 15 minutes, then spoon out 3 tablespoons of the stock into a small bowl and mix with the almond butter to make a smooth paste.

Add the spring greens to the soup and simmer for 5 minutes. Remove the chicken, shred into bite-sized pieces and return to the soup. Add the almond paste to the soup and stir well to combine, season to taste and serve in bowls with a wedge of lime on the side and a sprinkling of chopped coriander (cilantro) over the top.

SPICED PEANUT PARSNIP SOUP

By pre-roasting the parsnips, you get a great caramelized flavour that gives this soup that extra flavour boost. It's like a high-five in a bowl.

SERVES
2

TAKES
45 MIN

- 3 parsnips, peeled and cut into small chunks
- 1 red onion, peeled and quartered
- 2 garlic cloves, peeled
- 2.5cm/1-inch piece of fresh ginger, peeled
- 2 tsp vegetable stock powder made up with 720ml/3 cups boiling water
- 2 Tbsp peanut butter
- 2 Tbsp coconut cream
- Pinch of chilli powder
- 2 tsp ground cumin
- 30g/1oz coriander (cilantro) leaves

FOR THE TOPPING
- 1 Tbsp coconut chips
- 1 Tbsp pumpkin seeds
- Finely grated zest of 1 orange
- 15g/½oz coriander (cilantro), chopped
- Sea salt and freshly ground black pepper

Preheat the oven to 200°C/400°F/gas mark 6. Line an ovenproof dish with baking parchment, add the parsnips, onion, garlic and ginger to the dish and roast for about 20 minutes.

Meanwhile, make the topping by stirring all the ingredients together to combine, seasoning with salt and pepper. Set aside.

Blitz the stock in a high-speed blender with the peanut butter, coconut cream, chilli powder, cumin and coriander (cilantro).

Remove the roasted veg from the oven and add to the blender. Power for 1–2 minutes on a medium-high speed until you get a smooth texture. You may wish to add more water, depending on your preferred thickness.

Serve in bowls, sprinkled with the topping.

EASY OVEN-BAKED CHICKEN SATAY

Easy peasy lemon lime squeezy... and a drizzle of peanut butter. Oh, and some honey too.

- ½ small bunch of coriander (cilantro)
- 1 red chilli
- 1 garlic clove, peeled
- 3 heaped Tbsp peanut butter
- 3 Tbsp dark soy sauce
- 2cm/¾-inch piece of fresh ginger, peeled and roughly chopped
- Finely grated zest of 2 limes and juice of 1
- 6–8 bone-in chicken thighs, with skin
- Olive oil, to drizzle
- 3 Tbsp runny honey
- 3 Tbsp toasted sesame seeds
- Sea salt and freshly ground black pepper

Preheat the oven to 190°C/375°F/gas mark 5.

Put the coriander (cilantro), stalks and all, in a food processor with the chilli, garlic, peanut butter, soy sauce, ginger and lime zest and juice. Add a couple of splashes of water and blitz to a smooth paste. Season with salt and pepper.

Place the chicken thighs in a roasting dish, spoon over half the satay mixture (reserve the rest in a bowl for serving) and coat the chicken well. Drizzle with a little olive oil and season. Bake in the oven for 30-40 minutes, until cooked through, drizzling the honey over the chicken 10 minutes before the end of cooking. Once golden, take out of the oven and sprinkle over the sesame seeds.

Serve on a bed of brown rice or with a fresh green salad, and the reserved satay sauce on the side.

PIP'S TOP TIP:

If you'd rather have skewers, then just replace the chicken thighs with 4 chicken breasts, cubed and threaded onto skewers (soak the skewers for 20 minutes if they are wooden). The skewers only need 15 minutes in the oven and then a further 5-8 minutes with the honey.

CHICKEN MOLE WITH RED CABBAGE SLAW

No, don't worry, no furry blind animals are required for this dish. Mole (pronounced MOH-lay) is a slow-cooked Mexican dish that uses nut butter and chocolate to make a rich stew.

- 2 Tbsp sunflower oil
- 8 bone-in chicken thighs
- 2 onions, roughly chopped
- 2 tsp ground cumin
- 2 tsp ground cinnamon
- 2 red chillies, roughly chopped
- 3 Tbsp water
- 3 garlic cloves, roughly chopped
- 2 Tbsp runny honey
- 50g/⅓ cup raisins
- 3 Tbsp nut butter (almond or peanut)
- 2 Tbsp chipotle paste
- 1 x 400g can/2 cups chopped tomatoes
- 30g/1oz dark chocolate
- Sea salt and black pepper

FOR THE RED CABBAGE SLAW
- 1 small red cabbage, thinly sliced
- 1 shallot, thinly sliced
- 3 Tbsp red wine vinegar
- Small drizzle of olive oil
- Dash of Sriracha

TO SERVE
- Sesame seeds
- 6 tortillas, warmed
- Sour cream
- 1 lime, cut into wedges

Heat the oil in a large flameproof casserole, remove the skins of the chicken thighs and season the flesh with salt and pepper. Add to the oil and brown on all sides, then remove to a plate. Add the onions and cook for 5 minutes, until softened. Add the cumin and cinnamon and cook for 1 minute.

Blitz the chillies, water, garlic, honey and raisins to a smooth paste in a food processor, then pour into the casserole. Add the nut butter, chipotle paste and tomatoes, then fill the empty tomato can with water and add to the casserole. Return the chicken to the casserole and season well. Bring to the boil then turn down the heat, cover and simmer, stirring occasionally, for 1 hour.

Remove the chicken pieces and shred the meat off the bones, using a fork on a chopping board. Return the shredded chicken to the sauce and grate in the chocolate. Cook, uncovered, for another 30 minutes.

Meanwhile, make the red cabbage slaw by mixing the cabbage and shallot together in a bowl. Add the vinegar, oil and Sriracha and mix to combine.

Sprinkle the mole with sesame seeds and serve with the warmed tortillas and red cabbage slaw, with sour cream and lime wedges on the side.

NUT BUTTER SLAW

Forget the watery stuff you get in the supermarket, homemade coleslaw is way more fresh, crunchy and flavoursome. This salad makes a perfect companion for a burger or quiche.

- 3 Tbsp almonds
- 2 carrots
- 1 fennel bulb
- 400g/14oz mixture of red and white cabbage
- ½ red onion, thinly sliced into rings
- 1 handful each of coriander (cilantro) and mint, finely chopped
- 4 Tbsp extra virgin olive oil
- 250ml/1 cup natural yoghurt
- 4 heaped Tbsp almond butter
- Juice of 1 lemon
- Sea salt and freshly ground black pepper

Toast the almonds in a dry frying pan for a few minutes, then roughly chop.

Use a food processor to grate the carrots. Next, finely shred the fennel and cabbage and mix with the grated carrots in a large bowl. Add the onion and herbs.

In a separate bowl, mix together the olive oil, yoghurt, almond butter, and some salt and pepper. Add the lemon juice and mix.

Pour the yoghurt dressing over the slaw mix, add the chopped almonds, reserving a few to sprinkle over the top, and mix.

CRUSHED CARROTS

Crushed, smashed, mashed or puréed, are all a good way of describing a lovely side dish that makes a great alternative to your usual boiled variety. The nutty flavour of the peanut butter really complements the sweet carrots.

- 800g/1¾lb carrots
- Knob of unsalted butter
- 1 Tbsp peanut butter
- 2 tsp runny honey
- Sea salt and freshly ground black pepper
- 2 Tbsp chopped parsley, to serve

Peel the carrots, halve lengthways and cut into 2cm/¾-inch pieces. Steam for 25 minutes, until tender, then transfer to a medium bowl and mash with a fork or potato masher. Stir in the butter, peanut butter and honey, and season with salt and pepper. Sprinkle with chopped parsley to serve.

HUMMUS

This hummus, or houmous, however you like to spell it, is a twist on the classic recipe, which typically uses tahini.

- 1 x 400g can/heaped 1½ cups chickpeas (garbanzo beans)
- 2 tsp peanut butter
- 1 garlic clove, crushed
- 2 Tbsp lemon juice
- 1 tsp crushed sea salt
- 4 Tbsp extra virgin olive oil, plus extra for drizzling
- Freshly ground black pepper
- Sprinkling of paprika, to serve

Drain the chickpeas (garbanzo beans), reserving 2 tablespoons of the liquid, and rinse with cold water. Set aside a few chickpeas for garnish, then add the rest to a food processor with the peanut butter, garlic, lemon juice, the 2 tablespoons chickpea liquid, sea salt and pepper to taste, and blitz, pouring the oil in as it processes, until it reaches a smooth consistency.

Serve in a bowl with a dusting of paprika, a drizzle of oil and the reserved whole chickpeas.

PEANUT SWEET POTATO GRATIN

A bit like a cuddle in a dish. You can either serve this as a side or as a meal in itself, with a fresh salad. Either way you'll feel all warm and fuzzy inside once finished.

- 6 medium-sized sweet potatoes
- 2 Tbsp sunflower oil
- 250ml/1 cup whipping cream
- 1 red chilli, finely chopped
- 6 garlic cloves, finely chopped
- 1 tsp sea salt
- 4 Tbsp peanut butter
- 2 tsp water

Preheat the oven to 200°C/400°F/gas mark 6. Wash the sweet potatoes, but do not peel them, then cut into discs 5mm/⅕-inch thick. Use a mandolin to do this if you have one; just mind your fingers.

In a large bowl, toss the sweet potato slices with 1 tablespoon of the oil, the cream, chilli, garlic and salt. Layer half the potato slices in a deep, medium-sized ovenproof dish, making sure they are lying flat.

Whisk the peanut butter with the remaining 1 tablespoon oil and a couple of teaspoons of water. Spread this over the layer of sweet potato, using a spatula to ensure it covers as much of the surface as possible. Add the rest of the sweet potato on top, with any remaining cream from the mixing bowl.

Cover the dish with foil and bake for 25 minutes, then remove the foil and bake for another 35 minutes. Place under a hot grill for 5 minutes to get the top extra crispy.

SMOKED BABA GANOUSH

Baba ganoush translated from Arabic means "pampered daddy". So the perfect dip to make on Father's Day, we say! Or any day when you feel the need to spoil your old man.

- 2 Tbsp olive oil
- 3 aubergines (eggplants)
- 2 Tbsp cashew butter
- 1 large garlic clove, chopped
- Juice of 1 lemon
- Sea salt and freshly ground black pepper

TO SERVE
- Sprinkling of chopped parsley
- Sprinkling of za'atar or toasted sesame seeds

Preheat the oven to 190°C/375°F/gas mark 5. Brush a baking sheet or roasting tin with 1 tablespoon of the oil.

Prick the aubergines (eggplants) a few times with a fork or the tip of a knife. Place each directly over a gas flame and char the skins, turning them so that they char evenly. (Alternatively, you can do this under a hot grill.)

When cool enough to handle, trim off the stem and cut each in half lengthways. Place cut side down on the prepared baking sheet and roast for 30–35 minutes, until very, very tender when pressed. Leave to cool to room temperature.

Once cool, use a spoon to scrape all the aubergine flesh into a food processor. Add the cashew butter, garlic, lemon juice and some salt and pepper, and blend to a purée.

Spoon into a bowl and drizzle with the remaining 1 tablespoon olive oil, parsley and some za'atar or toasted sesame seeds.

RE-USE REDUCE RECYCLE

We're halfway through this book now. And unless you've made your own nut butter (amazing, well done) you might have a few empty jars knocking about.

There are lots of different ways that you can upcycle your jars. Our favourite is to create this handy bird feeder. Not only do you get to enjoy your favourite nut butter, but the birds get a tasty meal too.

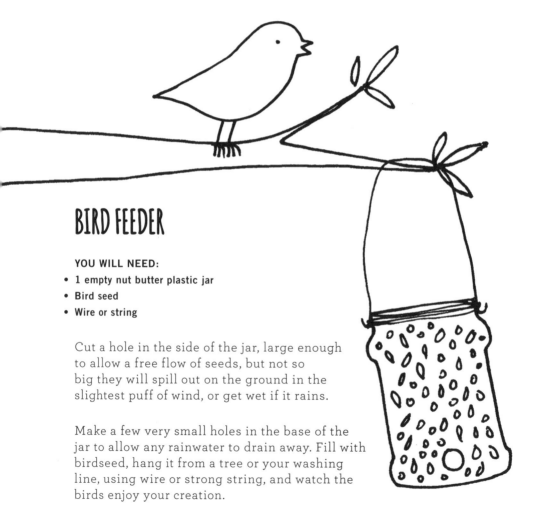

BIRD FEEDER

YOU WILL NEED:
- 1 empty nut butter plastic jar
- Bird seed
- Wire or string

Cut a hole in the side of the jar, large enough to allow a free flow of seeds, but not so big they will spill out on the ground in the slightest puff of wind, or get wet if it rains.

Make a few very small holes in the base of the jar to allow any rainwater to drain away. Fill with birdseed, hang it from a tree or your washing line, using wire or strong string, and watch the birds enjoy your creation.

DRESSINGS

Nut butter is a great addition to most dressings as it helps to bind the oils together and adds a great nutty depth to the dressing. The easiest way to make a dressing is using an old jam jar as a shaker to mix it up.

1. HONEY CASHEW DRESSING

A sweet little dressing to tart up any salad. It's especially good over a French bean salad.

 1 LARGE SALAD

 TAKES 5 MIN

- 2 Tbsp cashew butter
- Finely grated zest and juice of ½ lemon
- 3 Tbsp water
- ½ tsp finely grated orange zest
- ½ Tbsp runny honey
- ½ garlic clove, very finely chopped
- 2 Tbsp olive oil
- Sea salt and freshly ground black pepper

Put all the ingredients, except the olive oil, into a small bowl. Then slowly pour in the olive oil, a little at a time, stirring as you go.

2. SRIRACHA PEANUT DRESSING

 1 LARGE SALAD — TAKES 3 MIN

Up the amount of hot sauce in this dressing if you think you can handle it!

- 2 Tbsp peanut butter
- 2 Tbsp rice vinegar
- 1 Tbsp dark soy sauce
- 1 tsp Sriracha

Add all the ingredients to a bowl and whisk until smooth.

3. THE DETOX KITCHEN ALMOND BUTTER DRESSING

Our friends over at The Detox Kitchen, who have a couple of amazing cafés in London, really know their way around the kitchen. They created this dressing using our almond butter, and it featured as a special on their menu for a month. It was so tasty, in fact, that we couldn't resist putting it in our book.

- Juice of 1 lime
- 1 Tbsp tamari
- 1 Tbsp groundnut oil
- 50g/3 Tbsp almond butter
- Pinch of salt

Place all the ingredients in a bowl and mix well with a spoon.

PIP'S TOP TIPS:

The dressing tastes great on a simple green salad but The Detox Kitchen do a wicked one using 1 finely sliced head of kale, ¼ white cabbage cut into thin strips, 2 finely chopped sun-dried tomatoes and 1 finely chopped spring onion (scallion). Simply assemble all the ingredients in a bowl, toss with the dressing and sprinkle with pumpkin seeds.

6

Baking
&
Dessert

CHOCOLATE AND ALMOND BANANA BREAD

Try slicing the cooked banana bread and placing between two pieces of greaseproof paper, then toast on both sides in a frying pan over a low-medium heat. It'll caramelize the banana and melt the chocolate in the bread, and make the house smell pretty amazing. Just writing about it makes us hungry.

MAKES 1 LOAF

TAKES 1 HR 20 MIN

- 100g/7 Tbsp melted butter, slightly cooled, plus extra for greasing
- 2 eggs, beaten
- 160g/¾ cup plus 1 Tbsp soft light brown sugar
- 120g/½ cup almond butter
- 250g/8¾oz ripe bananas (peeled weight), mashed with a fork
- 180g/1⅓ cups plain (all-purpose) flour
- 2½ tsp baking powder
- 1 tsp salt
- 100g/3½oz dark chocolate, roughly chopped

Preheat the oven to 160°C/325°F/gas mark 3. Butter a 450g/1lb loaf tin, then line with baking parchment.

Put the melted butter in a bowl and add the eggs and sugar. Mix together to a smooth paste, then beat in the almond butter and mashed bananas.

Sift the flour, baking powder and salt into a separate bowl, then gradually fold the flour mixture into the banana mixture. Stir in the chopped chocolate and spoon the mixture into the prepared loaf tin.

Bake in the oven for 55 minutes, or until a wooden skewer inserted into the centre comes out clean. Leave to cool in the tin on a wire rack for about 10 minutes, then turn out of the tin onto the wire rack and leave to cool some more.

THREE NUT BROWNIE

Not one, not two, but THREE nuts in this brownie. That's right people, we've gone nuts for nuts in this recipe.

MAKES
16

TAKES
45 MIN

- 250g/1 cup plus 2 Tbsp unsalted butter, diced, plus extra for greasing
- 250g/8¾oz dark chocolate (at least 70% cocoa solids), broken into pieces
- 4 eggs
- 250g/1¼ cups golden caster (superfine) sugar
- 2 tsp vanilla extract
- 140g/1 cup plus 1 Tbsp plain (all-purpose) flour
- 50g/½ cup cocoa powder
- 75g/¾ cup chopped hazelnuts
- 40g/3 Tbsp almond butter
- 40g/3 Tbsp peanut butter
- ¾ tsp flaky sea salt

Preheat the oven to 180°C/350°F/gas mark 4. Butter a 20 x 20cm/8 x 8-inch square baking tin and line with baking parchment.

Put the diced butter and chocolate in a heatproof bowl set over a pan of simmering water, making sure the base of the bowl is not touching the water. When completely melted and combined, take the bowl off the heat and set aside to cool a little.

In a separate bowl, whisk the eggs and sugar together, using an electric whisk, until pale and light. Add the egg mixture to the chocolate mixture, along with the vanilla extract. Sift in the flour and cocoa and fold them in until combined. Stir in the chopped hazelnuts.

Tip the batter into the prepared tin, then blob dots of the nut butters evenly across the top. Using a wooden kebab stick, run through the middle of each nut butter blob to create a marbled effect. Sprinkle the sea salt over the top and bake for 25 minutes, until cracked on top but still gooey in the middle, bearing in mind that they will continue to firm up as they cool.

Leave to cool in the tin for 15 minutes, then remove to a wire rack to cool completely. Cut into 16 generous-sized brownies.

BAKED PEANUT BUTTER AND RASPBERRY CHEESECAKE

A PBJ sandwich, but in a cheesecake form. Winner.

- A little butter, for greasing
- 900g/scant 4 cups cream cheese
- 190g/1 cup caster (superfine) sugar
- 1 tsp vanilla extract
- 4 eggs
- 75g/5 Tbsp peanut butter
- 300g/10½oz raspberries
- Icing (confectioners') sugar, for dusting

FOR THE BASE
- 140g/1 cup plus 1 Tbsp plain (all-purpose) flour
- ¼ tsp baking powder
- 50g/¼ cup caster (superfine) sugar
- 50g/3½ Tbsp unsalted butter
- 1 egg yolk

Preheat the oven to 150°C/300°F/gas mark 2. Butter a 25cm/10-inch round springform cake tin and line with baking parchment.

To make the base, put the flour, baking powder, sugar and butter into the bowl of a free-standing mixer and mix until sandy. Add the egg yolk and mix in. Tip the mixture into the prepared tin and press it firmly into the base. Bake in the oven for 20–25 minutes, then set aside to cool.

Put the cream cheese, sugar and vanilla into the cleaned out bowl of the mixer and slowly combine until glossy. Add the eggs one at a time, still mixing, scraping the ingredients down the sides of the bowl from time to time. When you have added all the eggs and the mixture is fluffy, fold in the peanut butter using a large spoon, then fold in 150g/5¼oz of the raspberries, making sure they are evenly dispersed.

Scrape the filling onto the cooled base and gently smooth it out evenly using a spatula. Wrap the outside of the tin with foil, place in a deep baking tray and add water to the tray to come two-thirds of the way up the tin. Bake for 30–40 minutes, until set but still with a slight wobble.

Carefully take the baking tray out of the oven and place the tin on a wire rack to cool. To serve, top with the remaining raspberries and dust with icing (confectioners') sugar.

NALMOND NICE CREAM

No, that's not a typo but our coined word for this delightful banana and almond healthy ice cream. It doesn't contain any cream, so not only is this ice cream super tasty but it also counts towards your five-a-day. Ice cream for breakfast, anyone?

SERVES 6

TAKES 1 HR 15 MIN PLUS PRE-FREEZING TIME

- 4 ripe bananas, chopped into 3cm/1¼-inch chunks and frozen
- 2 Tbsp almond milk
- 3 Tbsp almond butter
- ½ tsp ground cinnamon

TO SERVE
- 1 Tbsp grated dark chocolate
- 1 Tbsp flaked (slivered) almonds

In a food processor, blend the frozen bananas and almond milk together to a smooth consistency. Add the almond butter and cinnamon, and blitz again.

Transfer to a freezer container and freeze for 1 hour, then serve sprinkled with grated chocolate and flaked (slivered) almonds.

CHILL-OUT TIME

ALMOND TOFFEE SAUCE

Eat on ice cream sundaes. Or pancakes. Or waffles. Brownies would work too. Apples also. Perhaps even on top of a fried banana. Basically anything and everything.

- 3 Tbsp light muscovado sugar
- 2 Tbsp coconut cream
- 2 Tbsp almond butter

Mix all the ingredients together in a pan over a low heat, stirring until the sugar has melted and the sauce is a dark coffee colour.

CHOCOLATE HAZELNUT SAUCE

Try to restrain yourself from drinking this through a straw. It's that goddamn good.

- 175ml/¾ cup double (heavy) cream
- 100g/3½oz milk chocolate, chopped
- 100g/scant ½ cup smooth hazelnut butter
- 3 Tbsp maple syrup

Put all the ingredients into a heavy-based pan and melt together over a low heat, stirring.

Simmer until the sauce is thick and creamy.

THE ULTIMATE (NUT BUTTER) KNICKERBOCKER GLORY

Knickerbocker glory. A brilliant word for an ice cream sundae but with a much better ring to it.

- 3 scoops of nut butter ice cream (see page 135) (vanilla ice cream will also work well)
- 75ml/⅓ cup chocolate hazelnut sauce (see page 127)
- 100g/3½oz salted pretzels

Take the ice cream out the freezer to soften.

Heat up the chocolate hazelnut sauce over a low heat, stirring occasionally. Crush the pretzels in a food bag, using a rolling pin, saving a whole one for decoration.

Using a nice tall ice cream glass, put a scoop of ice cream in the bottom, then sprinkle with pretzel crumbs and then the chocolate hazelnut sauce. Repeat and finish with the final scoop of ice cream. Top with the reserved whole pretzel and serve.

VEGAN ALMOND FUDGE

Make these little guys. Put them in the freezer. Forget about them. Then rediscover them when foraging in the freezer for some peas, and feel like you've won the lottery.

MAKES 21 CUBES

TAKES 15 MIN — PLUS FREEZING TIME

- A little vegetable oil, for greasing
- 50g/1¾oz coconut oil
- 60ml/¼ cup maple syrup
- ¼ tsp sea salt
- 350g/1½ cups almond butter

Lightly grease a 450g/1lb loaf tin and line with baking parchment so it overhangs the long sides.

Melt the coconut oil, maple syrup and salt in a small saucepan over a low heat.

Spoon the almond butter into a large mixing bowl, then slowly pour in the melted ingredients, stirring as you go. Keep stirring until completely smooth and combined. Don't worry – at this stage it will be quite runny.

Spoon the mixture into the prepared tin and smooth the top with a spatula. Place the tin, uncovered, on a flat surface in the freezer and freeze for at least 1 hour, or until solid.

Take out the freezer and, using the edges of the baking parchment, pull the whole slab out of the tin. Cut the slab into 21 cubes (lengthways into 3, then across into 7) then place in an airtight container and store in the freezer. They will keep in the freezer for up to 3 months.

HAZELNUT, APPLE AND PEAR CRUMBLE

Who doesn't love a crumble? Tell us, and we'll go and tell them what's what.

SERVES
4–5

TAKES
50 MIN

- 500g/1lb 1oz Bramley apples
- 500g/1lb 1oz pears
- Finely grated zest and juice of 1 orange
- About 50g/¼ cup caster (superfine) sugar (depending on the sweetness of the fruit, you may need more)

FOR THE TOPPING
- 75g plain/½ cup plus 1 Tbsp (all-purpose) flour
- 50g/⅓ cup toasted hazelnuts, roughly chopped
- 15g/2 Tbsp rolled oats
- 45g/3½ Tbsp soft brown sugar (light or dark)
- ½ tsp ground cinnamon
- 65g/scant ⅓ cup hazelnut butter

Preheat the oven to 160°C/325°F/gas mark 3.

Peel, core and slice the apples and pears, then spread out in a shallow ovenproof dish. Sprinkle over the orange zest and juice, then sprinkle the sugar evenly over the top.

For the topping, put the flour, chopped hazelnuts, oats, sugar, cinnamon and hazelnut butter into a food processor and pulse until the mixture resembles breadcrumbs. Sprinkle evenly over the fruit and bake in the oven for 30 minutes, or until the topping is pale golden.

Serve warm with a dollop of yoghurt or cream.

NUT BUTTER ICE CREAM

Any nut butter works in this recipe. Peanut packs a bit more a punch, whilst others are more subtle in flavour.

SERVES 4

TAKES 1 HR PLUS FREEZING TIME

- 100g/scant ½ cup peanut or almond butter (other nut butters also work)
- 300ml/1¼ cups whole milk
- 85g/scant ½ cup caster (superfine) sugar
- 2 egg yolks
- 300ml/1¼ cups whipping cream

Whisk the peanut or almond butter and milk together in a heavy-based saucepan over a low heat.

Meanwhile, in a bowl, beat the sugar and egg yolks together until pale and thick. Stir the hot milk and nut butter mixture into the eggs and sugar to combine, then pour the mixture back into the pan.

Cook, stirring all the time over a low heat and making sure it doesn't boil, until it starts to thicken and then coats the back of a wooden spoon. Pour into a bowl and leave to cool, then chill.

Whip the cream to soft peaks and fold into the chilled custard mixture. Churn in an ice-cream machine, according to the manufacturer's instructions.

BRAIN FREEZE TIME

ALMOND AFFOGATO

A great way to combine dessert with after-dinner coffee.
Two birds, one stone.

- 1 scoop of nut butter ice cream, made with almond butter (see page 135)
- 50ml/scant ¼ cup (1 large shot) freshly made espresso

Add the ice cream scoop to a small bowl or coffee cup and pour the hot espresso over the top. Serve while warm and melting.

PIP'S TOP TIP:

If serving this to a few people, you can prepare it in advance by serving the ice cream into cups and keeping them in the freezer. This means that when you come to serve you will have a cold cup so the ice cream won't melt straight away when you pour on the espresso, giving you time to get to the dinner party table in time!

ESPRESSO ALMOND CHOCOLATE TART

Looking for a show-stopper of a dessert? Then look no further.

SERVES 8–10

TAKES 1 HOUR

PLUS SETTING TIME

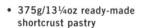

- 375g/13¼oz ready-made shortcrust pastry
- 300ml/1¼ cups double (heavy) cream
- 75g/½ cup caster (superfine) sugar
- ½ tsp sea salt
- 50g/3½ Tbsp unsalted butter
- 200g/7oz dark chocolate (at least 70% cocoa solids), broken into chunks
- 2 tsp instant espresso
- 50ml/scant ¼ cup whole milk
- 6 tsp almond butter

Preheat the oven to 180°C/350°F/gas mark 4.

Roll out the pastry to 5mm/⅕-inch thick and use to line a 25cm/10-inch tart tin. Line with baking parchment, fill with baking beans or dry rice and blind bake in the oven for 10–15 minutes. Remove the beans and parchment and bake for another 15 minutes, until it starts to turn golden brown. Set aside to cool.

Bring the cream, sugar and salt to the boil in a heavy-based pan. As soon as the cream starts to boil vigorously, take off the heat and add the butter, chocolate and espresso powder.

When the chocolate and butter have melted, stir in the milk, transfer the mixture to a jug and pour into the cooled pastry case, filling it right up to the rim but being careful not to let it overflow.

Using a small teaspoon, place small blobs of almond butter in rows running from one side of the pastry to the other. Using a wooden skewer, run through each blob to make a marbled effect.

Leave to set for a few hours. Enjoy cold with some double (heavy) cream.

FRUIT AND NUT FROZEN YOGHURT ICE LOLLIES

Lollies are very simple to make but you can really get creative with different flavours. This recipe is a suggestion, but get experimenting with your own ideas, as so many different flavours work!

  PLUS FREEZING TIME

- 1 medium, ripe banana
- 240ml/1 cup Greek yoghurt
- 1 tsp lemon juice
- 3 Tbsp honey
- 3 Tbsp nut butter
- Fresh summer berries

Purée the banana, yoghurt, lemon juice, honey and nut butter together in a food processor, until smooth. In a bowl, gently mash the summer berries to release some of their juices, but leaving the berries partially formed. Roughly fold the berries into the puréed mixture, making a ripple effect.

Divide the mixture between ice lolly moulds, stopping about 2.5cm/1 inch from top. If you don't have ice lolly moulds, use small paper cups.

Place a sheet of cling film over the top of each, cut a slit in the centre and add a wooden ice lolly stick. Freeze until firm, about 6 hours.

To release the ice lollies, dip the moulds briefly in hot water.

10 WAYS TO EXERCISE WITHOUT EVEN REALISING IT

Ok, so some of those baking recipes were pretty indulgent. And let's face it, sometimes one slice of cake just doesn't cut it. So here are some of the ways to exercise without even realizing it.

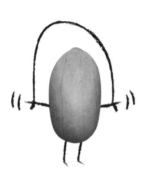

1. SKIPPING

Not just for the school playground, no sir. Skipping is actually a killer calorie-burner. Grab a rope with friends and see who can keep going for the longest.

2. GIVING THE HOUSE A ONCE OVER

Granted, no one likes cleaning the house but, believe it or not, whipping round the living room with your vacuum cleaner is in fact exercise.

3. DEN BUILDING

Ok, so your housemates might think you're strange but if you've never built a den then you are seriously missing out. A favourite weekend pastime of big and little kids alike – the bigger the better. And yes, once you've built your den it is acceptable to spend the rest of the day curled up inside watching films.

4. GETTING OFF THE METRO/BUS ONE STOP EARLIER

One of the easiest ways to get a bit of exercise in, and you don't even have to put your gym lycra on! Just get off the bus one stop earlier and stretch your legs a bit. In fact if you try to take a slightly different route every day, you'll even start discovering new places!

6. LEARNING A MARTIAL ART

Martial arts are not only great for making you feel like a superhero, they can be a pretty handy skill to have if you ever find yourself in a sticky situation.

5. HAVING A BOOGIE

Having a boogie, as your dad might say, is a sure-fire way to get your heart rate up and put a smile on your face.

7. DIY

Every household has that list on the fridge of 'odd jobs to do' that has been there for about two years, right? Well, putting up those shelves or assembling the flat-pack furniture is actually a pretty good workout.

8. CYCLING TO WORK

Not much to explain here.
Good for you AND saves you money.

6. YOGA

No, you don't have to be able to put your leg behind your head to become yogi. There are lots of different types of yoga, so try a few out and see which one you like the best. Not only does yoga encourage you to take a second out of your day and just breaaaaaaathe, it is also great for increasing strength and flexibility.

10. GARDENING

Give your garden a bit of love and spend some time mowing the grass or planting some veggies; you'll be surprised how hard it is. If you don't have your own garden, check out what's going on in your local community and if there is anything you can help with.

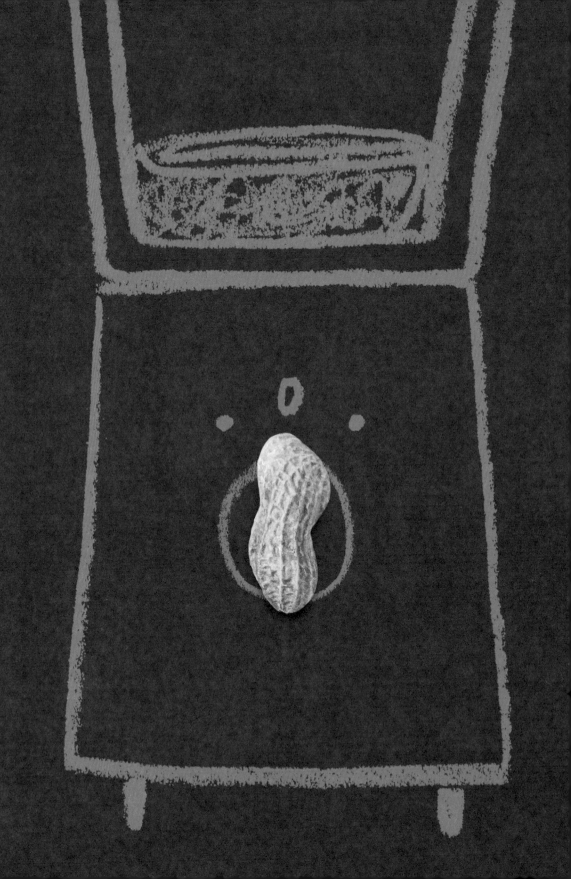

7.DRINKS

EVERYDAY SMOOTHIES

Smoothies. A great way to get one or two of your five-a-day down you (not literally, we hope) without even trying. Adding a tablespoon of nut butter here and there adds a bit of protein to your drink to help keep you going till lunch (or at least till elevenses – see snacks section on pages 66–83).

Below are three simple yet tasty recipes that you can easily make before work or school. Granted, you'll probably wake up the whole house in the process but they probably should be up anyway, right?

1. VIRGIN PIÑA COLADA SMOOTHIE

TAKES 5–15 MIN

A tropical one to take you back to your summer holiday, *sigh*.

- 100g/3½oz pineapple, peeled, cored and chopped
- 1 ripe banana, cut into chunks
- 2 large handfuls of spinach
- 2 Tbsp coconut almond butter
- 200ml/¾ cup coconut water
- 1 Tbsp chia seeds (optional)

Put all the ingredients into a blender and blitz until nice and smooth. If you have included the chia seeds, leave the smoothie for 10 minutes or so before drinking, to give the seeds time to absorb some of the liquid and swell up.

2. BIG BREAKFAST SMOOTHIE

We tried creating a smoothie out of eggs, bacon, sausage and beans, but it just didn't work out so well. Instead we settled for this combination.

- 1 frozen banana
- ½ apple, skin on, cored
- 2 medjool dates, pitted
- 50g/½ cup rolled oats
- 2 Tbsp peanut butter
- 1 Tbsp flaxseeds
- ¼ tsp ground cinnamon
- 100ml/scant ½ cup water
- Squeeze of lemon juice

Blitz all the ingredients together in a blender until smooth. If you find that it's a little too thick for your liking, just add a touch more water and blend some more.

PIP'S TOP TIP:

If you aren't using a frozen banana, swap the water for ice.

KEEPS YOU GOING UNTIL LUNCH

3. BERRIES 'N' CREAM

There's actually no cream in this recipe, but the combo of cashew butter and oats makes this smoothie really thick and creamy.

- 1 medium banana, cut into chunks
- 50g/1¾oz frozen berries
- 15g/1½ Tbsp rolled oats
- 2 Tbsp cashew butter
- 150ml/⅔ cup unsweetened almond milk
- 1 tsp coconut oil
- 1 tsp coconut nectar

Blitz everything together in a blender until fully combined.

PIP'S TOP TIP:

Go forage for berries in the frozen aisle of the supermarket. They're great when berries are out of season and because they're frozen they'll make your smoothie extra cold.

SUPERFOOD SMOOTHIES

Smoothies that make you feel super.

1. ALMOND BUTTER MATCHA SMOOTHIE

Matcha has some pretty awesome health benefits: helps you concentrate; boosts memory; increases energy levels. When we say super, this smoothie really is just that.

- 1 medium pear, cored and diced
- 2 handfuls of baby spinach leaves
- 1½ Tbsp almond butter
- 1 tsp matcha powder
- 1 tsp agave nectar
- 180ml/¾ cup unsweetened almond milk
- Small knob of fresh ginger, grated

Put all the ingredients in a blender and blitz until smooth.

PIP'S TOP TIP:

Adjust the matcha and ginger quantities to taste, as both are quite punchy flavours.

TEA-RIFFIC TASTE

2. SPICED CASHEW SWEET POTATO SMOOTHIE

Sweet potato in a smoothie. Yes. It does work. In fact not only does the sweet potato make the smoothie a lovely texture, it really keeps you going right up until lunch.

- 150g/5¼oz sweet potato
- ½ mango
- 1cm/⅓-inch piece of fresh ginger, grated
- 2 Tbsp cashew butter
- 2 tsp maple syrup
- 200ml/¾ cup unsweetened almond milk
- 2 ice cubes

Microwave the sweet potato until it is soft throughout. Scrape the flesh out of the skin and leave to cool.

Put into a blender with the remaining ingredients and blitz until smooth.

3. SUPER PINK SUNFLOWER COCOA SMOOTHIE

The colour of this smoothie is reason enough to make it.

- 1 medium beetroot, peeled and roughly chopped
- 100g/3½oz raspberries
- 1 Tbsp dark chocolate chips
- 3 Tbsp almond butter
- 1 Tbsp honey
- 150ml/⅔ cup apple juice

Put all the ingredients into a blender and blitz until frothy.

4. ALMOND KALE SMOOTHIE

Dates, check. Almond butter, check. Kale, check. Coconut water, check. All the trendiest ingredients in one pint-sized glass.

1

TAKES
2 MIN

- ½ orange, peeled
- 2 medjool dates, pitted
- 50g/3 Tbsp almond butter
- 1 banana, roughly chopped
- 60g/2oz kale leaves
- 100ml/scant ½ cup coconut water
- 5 ice cubes

Put all the ingredients in a blender and blitz until smooth.

ALMOND BUTTER
MATCHA SMOOTHIE

SUPER PINK SUNFLOWER
COCOA SMOOTHIE

153

FROTHY DRINKS

1. PEANUT BUTTER, BANANA AND HONEY MILKSHAKE

Peanut butter smoothie. It's a classic. Best served with lots of colourful straws so you slurp your way right to the bottom.

- 4 scoops of vanilla ice cream
- 4 Tbsp peanut butter
- 285ml/1¼ cups whole milk
- 2 bananas, cut into chunks
- 2 Tbsp runny honey

Blitz all the ingredients together in a blender until thick and creamy.

MAGIC MILKSHAKE

2. PEANUT BUTTER HOT CHOCOLATE

Just picture it. Slippers on. Log fire going. Your favourite film on the TV. All that, and a peanut butter hot chocolate in your hand.

- 175ml/¾ cup water
- 3 Tbsp cocoa powder, plus extra to finish
- 600ml/2½ cups milk
- 100g/3½oz dark chocolate, (at least 70% cocoa solids) broken into pieces
- 1½ Tbsp demerara sugar
- 3 Tbsp peanut butter
- Whipped cream, to finish

Bring the water to a simmer in a saucepan over a medium-high heat. Whisk in the cocoa powder until no lumps remain, then add the milk and bring back to a simmer. Whisk in the chocolate and sugar until the mixture is smooth and creamy and the chocolate is melted. It takes about 5 minutes. Add the peanut butter and stir well to mix.

Divide the hot chocolate between 2 mugs, top with whipped cream and dust with cocoa powder.

(Photograph opposite.)

3. ESPRESSO ALMOND FRAPPE

Save yourself a few pennies and make your own frappe.

- 3 medium scoops of vanilla ice cream
- 5 ice cubes
- 300ml/1¼ cups chilled brewed coffee
- 200ml/¾ cup whole milk
- 2 Tbsp almond butter

TO FINISH
- Whipped cream
- Cocoa powder

Blitz all the frappe ingredients together in a blender. Pour into glasses and top with whipped cream and a dusting of cocoa powder.

INDEX

159

THANKS

I had a few squirrels help me along the way while creating this book and so a big thank you goes to...

- All those that suggested recipes. Creating 70 recipes was harder than it looked so thanks to all those who gave me hints and tips. Special mentions to Lily Simpson from The Detox Kitchen who created, put on the menu and then allowed me their salad recipe for my very own book. Plus Willow Conway for her tried and tested granola recipe.

- Those that helped me research some of the chapters; in particular Ella who helped compile the smoothies and Nika who now is the font of all knowledge on nuts and seeds.

- My mum and dad who (un)willingly opened their kitchen to allow me to test out all the recipes and for putting up with all the mess that then ensued. And the rest of my family who were the guinea pigs.

- Jessie for thoroughly checking the book and made sure no mistakes slipped through the net.

- Adrian for taking some stunning pictures and really bringing the recipes to life.

- The fabulous team at B&B Studio, in particular Shaun, George, Kerry C, Georgie, Mathilde, and Kerry B for masterminding the brilliant design of the book.

- Quadrille for putting the whole book together, shout outs to Nikki and Gemma for their design skills, and Sarah and Helen for not only their help the whole way through but also for having the idea to commission the book in the first place!

- And finally, but most importantly, a big thank you to all those people who buy Pip & Nut. This book wouldn't be here without you.